W9-CBN-873

Praise for *Building Team Power*

"When I think of teamwork, Tom Kayser's name comes immediately to mind. As my colleague at Xerox, he facilitated team building workshops for my leadership teams that instilled in us the principles of how an effective team communicates and works. This book by Tom shares what we benefited from working with him in person. A must read!"

—Michael J. Lo, Ph.D., Vice President, Worldwide Marketing, Business Solutions and Services Group, Eastman Kodak Company

"*Building Team Power* systematically breaks down the barriers that have kept groups from being effective. It promises to be a useful reference for diagnosing and healing dysfunctional work groups of all types both today and tomorrow."

—Bob Nelson, author of the million-plus bestseller *1001 Ways to Reward Employees*

BUILDING TEAM POWER

How to Unleash the Collaborative Genius of Teams for Increased Engagement, Productivity, and Results

Second Edition

THOMAS A. KAYSER

NEW YORK CHICAGO SAN FRANCISCO LISBON LONDON
MADRID MEXICO CITY MILAN NEW DELHI SAN JUAN
SEOUL SINGAPORE SYDNEY TORONTO

The McGraw-Hill Companies

1 2 3 4 5 6 7 8 9 0 QFR/QFR 1 9 8 7 6 5 4 3 2 1 0

ISBN: 978-0-07-174674-8
MHID: 0-07-174674-9

Library of Congress Cataloging-in-Publication Data

Kayser, Thomas A., 1942-
Building team power : how to unleash the collaborative genius of teams for increased engagement, productivity, and results / by Thomas Kayser. — 2nd ed.
p. cm.
Includes index.
ISBN 978-0-07-174674-8 (alk. paper)
1. Teams in the workplace. I. Title.
HD66.K388 2011
658.4′022—dc22

2010032043

McGraw-Hill books are available at special quantity discounts to use as premiums and sales promotions, or for use in corporate training programs. To contact a representative, please visit the Contact Us pages at www.mhprofessional.com.

This book is printed on acid-free paper.

CONTENTS

ACKNOWLEDGMENTS

The power of collaboration was the key ingredient that made this book a reality. The biggest collaborators were all managerial and individual contributor colleagues I worked with at the Xerox Corporation for 30 years and my consulting clients since then. All the tools, processes, and structures written about here have been battle-tested by me many times over many years. Every encounter with a Xerox team or consulting client has been a hands-on learning experience for me to create, practice, refine, and hone the tools and processes written about here. Minus these collaborative opportunities to learn and experiment, I could never have written this book.

I am grateful for the many hours of help my wife, Carol, gave to this project. She has worked with much of this material in service and educational settings—district superintendents and their staffs, as well as school principals and their teams. Her work validated that what is written here has appeal beyond business and industrial situations.

A special debt of gratitude goes to Gary Krebs, vice president and group publisher, for the McGraw-Hill Companies, Inc., because Gary was the vocal champion inside McGraw-Hill calling out for this book. Gary was in my corner all the way. Thank you to Stephanie Frerich, my editorial director, for her management of this project through the various publishing steps inside McGraw-Hill.

And lastly, thanks to my son Chad, his wife Megan, and their miniature schnauzer Otto for your visits to pull me away from my computer and remind me that there is more to life than fretting over a manuscript "when you're chopping but the chips aren't flying."

FOREWORD

"Coming together is a beginning. Keeping together is progress. Working together is success."

Henry Ford

I first got to know Tom Kayser at Xerox. I had just completed the Executive Development MBA Program at the University of Rochester's Simon School. As a developing executive within the Xerox Human Resources organization and with my new degree in hand, my senior management sponsors thought that it would be a good idea (and developmental for me) if I headed up the Organizational Development function for the Business Products and Systems Group. I wasn't convinced that it was a good idea career-wise, but I agreed to give it a go.

Tom was the senior member of my new team of internal consultants. One of the first things he did was take me aside and teach me how to manage the group in a collaborative way. This was sorely needed because I was a relatively inexperienced manager, and my new team taught and consulted on management best practices across the company.

Over the next several years we laid the foundation that enabled our business group to win the coveted Malcolm Baldrige Award. We fostered and worked with hundreds of quality improvement and problem-solving teams teaching them team-oriented group processes, meeting skills and problem solving. We trained and enabled facilitators and employee networks. We helped plan and stage large global Teamwork best-practices events that drew teams from around the world. We coached leaders and taught them how to be more collaborative. We changed the corporate culture. I can't think of a better laboratory where Tom was able to test his thinking and hone his ideas.

Little did I know that this assignment would significantly alter my future. I discovered that I wanted to try consulting. And after leaving Xerox, I went on to become CEO of four different consulting and training companies doing business around the world. I learned a lot along the way in those endeavors, but every day I still use the tried-and-true ideas and skills that Tom taught me when we worked together. And we've come full circle. Tom and I are working together again as consultants, helping one of our large, well-known clients to build a more collaborative culture.

One of the things that I really appreciate about Tom is his knowledge and his respect for learning. Tom does an extraordinary amount of research, and he appreciates and recognizes the work of others who have gone before us. He reveres the seminal thinkers and appreciates the timelessness of their ideas and discoveries. He has an incredible grasp of who has written what and how that ties into the later work of someone else. Tom knows what he's talking about and carefully studies and tests before he recommends.

Tom has a gift for boiling complex things down to the essence of what's most important. He doesn't hit you with a lot buzzwords and consulting-speak. In fact, we had a phrase that we coined back at Xerox and we still use it today. Most of the stuff we do operates along *"the cutting-edge of common sense."* Tom strives for complete clarity, and you will see that in his writing style. He is a teacher at heart and does an artful job of crafting his message so that the reader can not just understand it, but learn it and use it.

Building Team Power addresses one of the biggest issues facing organizations today . . . *collaboration*. As Tom lays out at the beginning of the book, today's work environment for many organizations is turbulent and resembles the whitewater in Class V rapids. Per the Conference Board, the number one issue cited by CEOs globally is *"excellence in execution."* Connecting the dots, I would offer that execution is done by people, and people must collaborate effectively in order to execute.

It's probably not too much of a stretch to imagine organizations like yours as rafts in the whitewater rapids. As you think about that, is everyone on your team paddling in the same direction? Does everyone in the raft even have a paddle? Do you know where you want to steer the raft? Does everyone understand that direction and buy into where you want to head? Do they know their roles and how they are supposed to paddle to best help the group steer the raft? This book will be a tremendous resource to you in answering those kinds of questions.

Our client, whom I mentioned above, is a good case in point about the strategic importance of collaboration. They had laid out their long-term vision and done a strategic analysis of what kind of culture will be required in the future. Their analysis determined that they needed to build a more collaborative culture across their decentralized divisions, which were spread across the country. Then, as part of their overall solution, they decided that our proposed approach was aligned with what they were trying to do.

Tom and I, along with a few other members of the Tailwind team, then helped them integrate collaboration into how they built their strategy, how they reached consensus on it across a decentralized structure, how they translated it for every executive and manager in the company, how they clarified the roles of each executive and manager in executing the strategy, how they aligned the company around it, how they develop their future leaders who will lead it, and how they run their meetings at every level. In their case, collaboration is not just one more thing to do; it's a key element of their strategy and a becoming part of their culture.

As my own professional journey has unfolded, I have become a big believer in universals. That is, the same problems or issues keep popping up in place after place. I'm not saying that one size fits all, but the root problems are similar and you will find them if you know where to look. And that's one of the key things I love about *Building Team Power.* Tom has taken a core set of those universals and put together a great resource for leaders and developing leaders. He has lent his expertise to the areas of Trust, Decision Making, Consensus Building, Conflict Management, Delegation, and Problem Solving—critical building blocks for leaders at every level. The chapters in *Building Team Power* unfold like a course in leadership at a top university. And Tom offers it up in a style where you can easily understand it and put it to use. All of this knowledge is centered on the art and science of *collaboration*.

Having personally had the opportunity to work with Tom, I am pleased that you have a similar opportunity to take advantage of the insights and approaches he shares in this book. Like me, I hope that your experience with Tom Kayser creates positive change in your own career and takes you to places that you can only imagine right now. Best wishes on your journey.

Ron Cox
Founder and CEO, Tailwind Consulting
Connecting People to Strategy
October 1, 2010

PREFACE

As I reached the front of the check-in line in Cleveland, Ohio, the airline representative asked me where I was going and I said, "Los Angeles." She asked if I was checking any bags, and I replied, "Three." I also stated, "I would like one bag checked to L.A., the second checked to Istanbul, Turkey, and the third to Auckland, New Zealand."

With an icy stare and a cutting tone to her voice she slapped her hand on the counter and declared, "Impossible! We can't do that."

"That's funny," I replied. "Last month you did it for me and I didn't even ask you to!"

We all have our favorite stories about lack of service and poor quality. In fact, baseball is no longer America's favorite pastime. Today our great pastime is recounting tales of airlines that miss schedules or lose luggage, computers that crash at the worst possible time, service personnel who are rude or unmotivated, new or nearly new automobiles that are recalled to fix serious safety issues, and toys that break after two minutes in the hands of a five-year-old.

While some stories may be a bit exaggerated, they all drive home the same point: suppliers of goods and services have a long way to go to meet the requirements of their customers, whether they are around the block or around the world. And one thing is clear: the traditional, hierarchical, command-and-control organization *is not* the mechanism for consistently meeting customer requirements in a dynamic, fickle, global marketplace. The pathway to your competitive success *is* building team power across your enterprise so you can unleash the untapped, latent collaborative genius of your people and your work teams. It's about building and sustaining a flexible span-of-support instead of erecting a rigid span-of-control.

Ken Blanchard's comments underscore my contention that a new way of doing business is mandatory and that this new way is built on collaboration.

> Gone are the proclamations from above, the 700-page policy manual and the "check your brains at the door" mentality toward workers of years past. Today's organizations must move fast to collaborate, and in the process find a way to come up with the right decisions more times than not. Companies need those employees who are closest to the action to be fully engaged in striving to bring their best thinking to the organization's most pressing needs. And those individuals must interact to develop even better decisions than any one employee could achieve on his or her own.
>
> The team is the perfect vehicle for having the dynamic of this interaction take place. The team is where the organization's needs and employee talents can converge. With a team, you can quickly bring a cross section of perspectives to bear on issues of pressing importance to the current and future well-being of the organization. You can quickly share relevant data for determining the most appropriate action given the situation at hand.[1]

THE NEED FOR THIS BOOK

After the publication and favorable reception accorded the three editions of my first book, *Mining Group Gold: How to Cash in on the Collaborative Brain Power of a Team for Innovation and Results* (New

York: McGraw-Hill, 1990, 1995, 2011), and the popularity of the first edition of *Building Team Power* (Burr Ridge, Illinois: Irwin, 1994), I knew I was on to something that went beyond casual interest.

My 30 years of experience at Xerox, leading and facilitating every possible type of organization effectiveness intervention; my debates with colleagues both inside and outside of Xerox; my interactions with clients as part of my current consulting practice; my discussions with my wife, Carol, who has worked with a number of school districts and industrial firms teaching and applying collaborative principles; along with my review of the business literature, have all validated the same thing time after time.

New managers most often do not possess collaborative know-how simply because of their managerial inexperience. Many experienced managers, on the other hand, lack collaborative skills and insights because they grew up in the centralized, command-and-control bureaucracy where large merit increases, promotions, bonuses, lavish offices, and other forms of reward and recognition went to those managers who were best at dictating and directing. In those order-giving organizations, nurturing collaboration and teamwork was not a highly valued managerial behavior. Because it was looked upon as being soft and weak, as well as a surefire career killer, skill development in building and facilitating collaborative partnerships was squashed.

> New and experienced managers alike lack the necessary mental maps, the keen understanding, and the crisp execution of the "how to" fundamentals for being a collaborative leader building collaborative partnerships within and across work teams.

While this medieval thinking still exists in many places today, you can no longer allow it to predominate and drive your company's culture. If your organization does not understand or believe in utilizing the power of collaboration—or dismisses it as some passing fad—you are riding on the same highway to failure as the airline that sends a customer's baggage to three different destinations!

Figuring out ways to do things right the first time, every time; creating products and services that delight your customers and distress your competitors; forming and implementing strategies and structures to capture and hold market share; and studying and reconfiguring work processes to slice costs and improve quality are huge tasks requiring the collaborative genius of your entire organization. It is irrelevant whether your establishment is in the public sector or private sector, is a producer of goods or services, or is for profit or is nonprofit—a culture of collaboration up, down, sideways, and diagonally within and across every corner of your organization is mandatory for success in today's global marketplace. You cannot afford to have it any other way.

WHAT YOU WILL GET OUT OF READING *BUILDING TEAM POWER*

This is a hands-on, how-to book. It is a roll-up-your-sleeves-and-get-your-hands-dirty book. *It is applications oriented all the way.* Don't look for complex psychological, sociological, or academic group theory models here. You won't find any. *Building Team Power* extends and expands the reach of my well-received *Mining Group Gold* book. This one digs into the crucial behaviors you need to understand and practice to be a collaborative leader. It takes you into the "how-tos" for building collaborative partnerships and facilitating teamwork within your own work group, across work groups, or in task forces, committees, problem-solving teams, executive councils, and the like.It is one thing to say, "We need more and better collaboration around here;" it is another thing to do it. This book fills a void because it shows you how to do it in terms you can understand, with skills you can actually put into practice! Skill improvement in building and facilitating collaborative partnerships and teamwork is what this book teaches.

After reading *Building Team Power,* you will, I hope, pick out a few collaborative leadership actions and get started by saying: "Hey, I can give these a try because I understand what I am supposed to do. Kayser's book taught me something."

MY APPROACH TO WRITING THIS BOOK

I have tried to eliminate the nonsense and boil down the subject matter so it is highly informative, yet fun to read. I've used a personal style. In other words, if I could sit down and talk with you for a day about building team power, this is what I would say to you. I have incorporated a core model into each chapter that drives its content. Besides the core model, each chapter contains personal experiences, stories, examples, case vignettes, and sidebars to make the book both enjoyable and instructive.

THE FOCUS OF BUILDING TEAM POWER

Building Team Power moves you along your excursion to learn how to unleash the collaborative genius of work teams by concentrating on seven skill areas. These skill areas equate to the seven chapters comprising Part II. To clarify the book's focus and to acquaint you with its content and flow, read through the Contents. The part titles provide the broad approach; the chapter titles and subtitles give you an outline of the content material.

IDEAS FOR MAKING THE TRANSITION TO COLLABORATIVE LEADER

The key to success in acquiring or improving your group facilitation skills is the old refrain: practice, practice, practice. There are no shortcuts. However, your skill development can be an organized endeavor using *Building Team Power* as your central resource. Your skill acquisition process involves six steps.

1. Read the book straight through to obtain a solid grounding in each chapter's content. As you read, highlight key points that are of particular interest.

2. As you finish reading Chapters Four through Ten, write a final set of notes on the "Notes Worksheet" and read the "Reflections" points that close out those chapters.
3. Choose several chapters that are the most important to you and review your highlighted points and your "Notes Worksheets." Add margin notes if you please.
4. Select the model from one of the chapters you want to practice first to improve either how you collaborate with a specific teammate or how you improve collaboration within your work team. There are suggested "Leader Application Activities" at the end of Chapters Four through Ten for going down either track.
5. Try out your chosen actions and assess where your efforts went well and where they did not go so well.
6. Make a new plan that preserves the positive aspects of your earlier effort while reducing or eliminating the negative aspects, then try again.

With each successive cycle, you and your teammates will make progress in your ability to pull together, and you will soon realize that everyone on the team—not just the manager or task force chairperson—shares responsibility for collaborative facilitation and building collaborative partnerships. Along these lines of making the transition from controlling to facilitating, Fran Rees offers some reinforcing advice.

> Leaders who are making the transition to a more facilitative approach will do well to remember three principles of change:
>
> - Change takes *time.*
> - Change is a *process,* not a decision.
> - Change requires plenty of *experience and practice* in the new way of doing things.
>
> Leaders seeking change must acknowledge the vast effort it takes and must not give up when it takes more time and practice than anticipated. Change is not on a switch, like a light that goes on and off, but is a process—sometimes a

very long one. Leaders who effectively implement change make sure their teams have plenty of opportunities to practice new ways of doing things while they go about their day-to-day activities. Also, ask yourself what strengths, support systems, and other resources you already have that can help you make these improvements. Plan to use these strengths to your advantage when you begin to make changes.[2]

SOME FINAL THOUGHTS

As you read and reflect on *Team Power,* keep the following in mind:

- All of these techniques are focused on helping you see and understand the world from the perspective of a collaborative leader.
- The points presented here are not inflexible prescriptions—cast in concrete—but rather, ideas for guiding you in initiating and maintaining collaborative relationships within your team or across several teams.
- Because these are only guidelines, you are encouraged initially to practice the techniques and processes as described. However, after gaining experience and confidence in their use, supplement and refine this foundation in order to extend your skills in facilitating teamwork and collaboration.

Finally, while I believe *Building Team Power* and *Mining Group Gold* make a wonderful pair of books, combining to provide an abundance of knowledge on the subjects of collaboration, collaborative leadership, team facilitation, team building, and the like, the two books can be read in any order.

> There is a Chinese proverb that states: "Even a journey of a thousand miles must begin with the first step." Now having read all the "stage-setting" front matter, let's take the first step of our journey to learn about the call for collaboration by moving on to Chapter One.

NOTES

1. T. Kayser, *Building Team Power: How to Unleash the Collaborative Genius of Work Teams* (New York: McGraw-Hill, Inc, 1994), vii.
2. F. Rees, *How to LEAD Work Teams: Facilitation Skills* (San Diego, California: Pfeiffer and Company, 1991), 43–45.

PART I

The Pressure for Collaborative Leadership and Partnerships in Organizations

CHAPTER ONE

THE CALL FOR COLLABORATIVE LEADERSHIP

Whitewater Global Markets and the Transformation Organizational Structures

CHAPTER OBJECTIVES

- To characterize global marketplace conditions of today
- To describe the characteristics of the traditional organizational model
- To contrast the traditional model with a transformed organizational model
- To detail five elements of the transformed model that are key to marketplace success
- To demonstrate that collaboration is what makes the transformed model "tick"

INTRODUCTION

The symphony orchestra is routinely held up as the classic example of collaboration in action. Ninety individuals playing their instruments in perfect harmony to produce music that sends a chill down your spine. Sitting in the theater, you have the opportunity to enjoy, firsthand, the results of that collaboration. However, collaboration rarely just emerges from an orchestra, or any work group for that matter. It requires someone to orchestrate it and others to pull together and follow that lead to make collaboration a reality. Initiating, nurturing, refining, and extending collaboration throughout a work team or between departments takes time, understanding, and patience on the part of everyone involved. Yet it can be fouled up in a heartbeat by anyone with misguided intentions. The following consultant's report on actions to improve the technical efficiency and productivity of a symphony orchestra vividly illustrates this point.

> All 12 violins are playing identical notes; this is unnecessary and wasteful duplication. The violin section can be cut drastically, saving considerable labor costs. The oboe players have absolutely nothing to do for long periods of time. They just sit in their chairs. Their number should be reduced. Compositions involving the oboe must be rewritten so that the work is spread out more evenly, thus eliminating costly "peaks" and "valleys" of oboe productivity.
>
> I noted a recurring repetition of certain musical passages. What useful purpose is served by repeating on horns what has already been produced by the strings? Were all such redundant passages eliminated, the concert time (2 hours) could easily be reduced to 40 minutes. This would also eliminate the need for a time-wasting intermission. Something should be done about the shocking obsolescence of equipment. The program notes informed me that the first violinist's instrument was several hundred years old. If normal depreciation schedules had been applied, the instrument's value would have been reduced to zero, and a more modern and efficient violin could have been purchased.[1]

An apocryphal story? Certainly. But, like many leaders in the workaday world, our change consultant had no understanding of, or appreciation for, the power of collaboration. In an attempt to make the members' work ever more simple and efficient in order to drive improvement in the orchestra's performance capability, this person was about to destroy the diversity and synergistic genius—the collaborative soul—of this team.

Any organization's success depends, first and foremost, on how well it is able to tap the creative and innovative potential of all its members, regardless of their level. While a massive amount of lip service is paid to the idea that "our people are our greatest resource," many approaches to management, cost accounting, productivity measurement, and technology actually view employees as variable costs to be controlled. Failing to recognize the collaborative value of their people, too many organizations encourage adherence to lock-step procedures and maintenance of the status quo. As you will soon see, given the fast-breaking, decentralized global marketplace, this is a death sentence.

THE BUSINESS ENVIRONMENT NOW AND AHEAD: CLASS V WHITEWATER TURBULENCE

Every kayaker, canoeist, or rafter knows well the classification system that describes the rivers and rapids they paddle ("Class I," easy, to "Class VI," unrunnable). "Class V" waters, the most turbulent of the runnable rivers and rapids, are long and contain more continuous features that cannot be avoided. Features include such things as: strong rapids, large waves, big holes, unpredictable currents, and dangerous obstructions requiring multiple maneuvers to get around or through; in addition, there is serious risk to those going overboard because others may not be able to help.

Class V whitewater rapids are a perfect metaphor for the turbulent business environment facing the vast majority of organizations today. Powerful technological, economic, and political forces are converging to create a new order in which nations all belong to a single global marketplace. Huge global markets for video-gaming systems come and go every few years. The Internet, satellites, and PDAs have integrated the world's financial markets to react on a split-second basis; when the financial earthquake hit Wall Street in the fall of 2008, London and Tokyo felt the tremendous aftershocks. More and more manufacturers are producing goods in Third World countries, in cities whose names we can't even pronounce. On the political front, an increasing number of nations are dismantling trade barriers and deregulating state-run industries—policies that pave the way for even more multinational interdependence in business.

So, as companies large and small, profit and nonprofit, manufacturing and service oriented, all journey farther down their rivers, navigating the waves, currents, rapids, and obstacles of their churning business environments, all they will find is more uncertainty, complexity, and interdependence as the decades roll by. To be able to keep paddling and maneuvering, the successful organizations have to rely less on centralized grand strategies—designed and dictated by senior management to the rest of the organization—and more on the collaborative abilities of managerial and employee work teams operating in nimble networks to decipher trends and react swiftly with appropriate responses in terms of decisions and actions.

In summary, the marketplace dynamics just discussed might best be characterized by the adage: the world is moving so fast these days, the person who says it can't be done is generally interrupted by someone doing it.

TRADITIONAL BUREAUCRACY: TOO RIGID AND SLOW FOR TODAY'S NEEDS

The business and economic success America enjoyed post-WWII came about because we took the structures and processes designed during the first decades of the twentieth century and ran them harder. In effect, we performed organizational surgery and injected organizational "steroids." Whenever economic conditions got tough, we restructured, delayered, downsized, cranked out costs, and made efforts to pump up perceived quality.

In essence, we took the same old organization and culture and just squeezed it, but in the overwhelming majority of instances *we didn't really transform anything.* We talked a good game *conceptually* about collaboration, synergy, and teamwork. Yet when push came to shove, over and over again we subscribed to the "birdcage approach to change."

The birdcage approach to change operates like this: The canaries are all sitting on their perches in the birdcage. You reach in and remove two canaries. The remaining canaries all flutter around the cage and come down on different perches. Voilà! You have downsized and reorganized your birdcage. But nothing was actually transformed. You simply have the same canaries—albeit fewer in number—sitting on different perches in the same birdcage, singing the same songs, and producing the same droppings as before.

The birdcage theory of change was used often during the 1980s, 1990s, and the first decade of this century. Thousands of managers' heads rolled, and those that remained sat on different perches. Thousands of industrial workforce members were laid off to show the world that American business was breaking away from its bloated, fat cat past. Slicing workers—and in most instances *talking about* the need to empower workers, to be more flexible, to be more risk taking, to be more innovative—has for too long been the change agenda for too many firms. However, the traditional bureaucracy, with the senior bosses as all-seeing, all-knowing, order givers, has remained largely intact. But this approach is no longer viable.

More and more organizations in the public and private sectors are coming to realize that the traditional bureaucracy has been pushed beyond its limits of effectiveness. It is an inadequate response to the new economic and market demands presented earlier.

Steven Dichter, former senior partner in the New York office of McKinsey and Company, foresaw in the 1990s exactly what is being taken seriously today on a broad scale.

> [Organizational] strategies are increasingly shifting from cost-and-volume-based sources of competitive advantage to focusing on increased value to the customer.... The command-and-control organization is under strain. Indeed, many businesses are finding that command-and-control principles now result in competitive disadvantage.

- *Cost.* Layers of management and unnecessary staff functions to communicate and control top management directives can no longer be afforded.
- *Slow Response.* Standardized procedures, together with inflexible roles and responsibilities, create an organization that does not readily sense and react to changes in customer needs or technologies.
- *Lack of Creativity and Initiative.* Narrowly defined tasks do not fully tap the potential of today's better educated employee.[2]

The first principle of Organization Theory 101 makes clear that optimal structures are situational. That is, they are influenced by each organization's location, market, environment, and history. The kind of structure that makes sense for a pet food manufacturer may not make sense for a software developer or a Broadway theater.

For organizations fortunate to navigate in calmer, "Class I or Class II" waters, traditional approaches work fine. The bureaucracy can function well in a relatively stable market environment, where problems can be identified by daily or weekly reports, analyzed by staff, presented to top management for their decision, and delegated to middle managers who transmit the decisions and coordinate the implementation of the workers. If you are in a stable, slow-growth mass market, using single-purpose machinery, semiskilled workers, and producing standardized, high-volume products (like chemicals, paper, lumber, electric power), the command-and-control bureaucracy can work for you. These and similar industries are less affected by the turbulent global market conditions described earlier. Still, competitive forces, even in stable industries, require those enterprises to be vigilant and to eliminate or rework overly restrictive policies, procedures, and control systems; to move final decision making down the hierarchy; and to search for ways to eliminate waste from their work processes.

Conversely, if you operate in industries facing the "Class V" whitewater business conditions—such as IT services, banking/financial services, software development, green energy, health care/pharma, mass media, autos, and the like—then you must make certain your bureaucracy is refined, loosened, and reshaped to be a world-class competitor. The reason is simple. A dynamic, uncertain, interdependent, and complex market environment overloads the bureaucracy's ability to be nimble, collaborative, and fast acting.

THE TRANSFORMED ENTERPRISE: NIMBLE AND FAST ACTING TO MEET TODAY'S NEEDS

In creating the transformed enterprise, today's organizations are not trying to destroy and eliminate the traditional model, but rather to shape and develop it differently so it can excel as a structure in coping with the chaotic conditions facing most businesses today.

As Table 1-1 shows, the elements of the traditional model still will exist even as the transformed enterprise evolves and takes root. The degrees of variation among organizations as they modify and refine the bureaucratic template to meet their business needs will be great. But the message is clear: while the basic elements underlying the traditional template will remain, they will be thoroughly reworked and transformed. Although it is still possible to operate in the traditional manner today, the intricacies and complexities of the marketplace are relegating this type of operation to the margins.

Table 1-1. The Traditional Model Transformed to Meet Today's Business Conditions

Traditional Bureaucratic Elements	Transformed Bureaucratic Elements
Hierarchy	**Some form of a hierarchy will exist:** You will have fewer layers, more teams, broader networks; but you still will have organizational levels—every person or unit will not report to one person.

Table 1-1. The Traditional Model Transformed to Meet Today's Business Conditions (continued)

Centralized Decision Making	**Decisions obviously will still be made:** while key decisions regarding strategic business direction, facilities planning, advertising, etc., remain fairly centralized, decisions concerning strategic implementation will be made locally, and many other product and customer service decisions will be pushed down the hierarchy closer to the action.
Personal Specialization	**People still will be specialists in some set of specified functions:** they may be skilled in four or five functions instead of one or two, but there is a limit to the number and variety of functions in which a person can be skilled.
Line Operations and Staff Support	**There will be some form of line operations and staff support:** the line operations may be built around more flexible teams working in parallel, doing many steps in a process rather than organized in rigid functions; and available staff support may be much smaller and have its role defined differently; but both will exist.
Rules, Regulations, Policies	**You are going to have some form of rules, regulations, and policies:** they may be more flexible, allowing leeway for different local conditions or situations, but they will exist to help set the culture—"the way we do business around here"—and to prevent anarchy.
Controls	**You will also have budget, headcount, or other control mechanisms:** they may be less rigid, there may be greater collaboration in

	setting and utilizing such mechanisms, but they will exist.
Functional Charters and Job Descriptions	**These will exist to some degree:** but those that do exist will be broadly written to provide a general framework to the business, and they will not be so rigidly interpreted as to create silos and an "it's not my job" mentality; also, they will be written to encourage teamwork and collaboration within and across functions.

INTERNAL OPERATIONS OF THE TRANSFORMED ENTERPRISE

If we pull together what is described on the right-hand side of Table 1-1, we can describe the transformed enterprise as being one that is: *flatter, more flexible, fast acting, team oriented, and customer driven.* Figure 1-1 graphically emphasizes these elements as five interconnected pieces.

Let's look briefly at each element to bring the whole operational picture into focus.

Figure 1-1. The Five Elements of the Transformed Enterprise

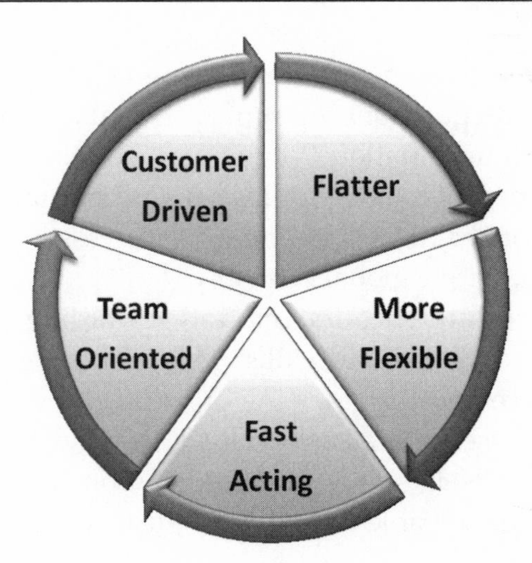

Flatter

Becoming *flatter* means stripping out unnecessary layers of the managerial hierarchy. *Unnecessary* refers to any person or function whose sole purpose is to act as a "mind guard"—sifting, sorting, interpreting, and censoring the information that flows between the hands-on people doing the work and the decision makers. The old hierarchical, pyramid-shaped management structure with its concurrent "mind guarding" is adept at grinding most decision making to a snail's pace. Companies today must be able to launch new products quickly and alter existing ones for big customers. The only solution is flatter companies, in which information flows quickly from top to bottom and back up again, compelling decisions to come fast.

However, the compression between the top and bottom that results from removing hierarchical levels has its dangers. Managers are thrust down closer to where the action is—closer to activities they do not readily know how to do or understand. The workers, on the other hand, who are used to carrying out directives, are now being asked to plan, schedule, solve problems, and make decisions. This restructuring places a premium on something that may have been in short supply under the old system: *teamwork and collaboration* up, down, and across all functions and levels of the hierarchy.

More Flexible

Increased *flexibility* means eliminating or redefining any procedures, rules, regulations, or approvals that are snarling decision making and information flow or perpetuating rigid work processes. In short, it means ripping out and burning your red tape! Organizational agility is essential for success in a fast-changing, highly competitive global economy.

Companies are encouraging and rewarding people to learn different knowledge and skills. This makes people a multifaceted resource and provides the organization with the ability to reconfigure itself more easily. This also builds learning into the job, which in itself is a source of motivation and reward for individuals. In the flexible organization, teams variously composed of first-line workers, managers, technical experts, suppliers, and customers *collaborate* to do a job and then disband, with everyone going off to the next assignment. Projects are being headed by *collaborative leaders* facilitating the activities of interdisciplinary talent. This feature aims for one thing: fluidity.

Fast Acting

Fast acting means speed. Having ideas is not enough without the ability to commercialize them before the competition does. Reducing time-to-market is critical because, as the saying goes: the early bird gets the worm.

In the global economy today, with fierce, world-class competitors lurking in every marketplace, the innovator has the edge. Early introduction of a product into the market gives the innovator several outstanding advantages: longer sales life, higher market share, higher margins because of premium pricing for being there first, and cost advantages from the manufacturing learning curve. Every organization must consider lost time as an irreplaceable resource. Either you put your own products out of business or your competitors will!

Creating teams aligned with the company's strategy, empowered with total project responsibility, and staffed with the smallest number of people having the necessary complementary skills to meet the project's defined goals, requires *collaboration* of the highest form. An emphasis on "chimney busting"—eliminating the communication barriers and endless reviews between functional departments so people can collaborate on a common project's success also drives speed.

Team Oriented

Team oriented means altering the classic hierarchical structure so it is no longer the sole or dominant determinant of organizational relationships. A diverse group of people—using their own creativity, innovation, judgment, intuition, and brain power—can do a better job in today's world of constant change than any set of formal procedures, methods, or controls administered by a remote, centralized management. And social science research is replete with studies on the power of synergy demonstrating that *the group social mind is far more than the sum of the individuals.*

When intra- and interteam collaborative relationships are supported and facilitated through words, actions, structures, and processes, the entire organization benefits through: an improved sharing of know-how; a more effective coordination of the flow of products and services from one unit to another; a greater willingness for resource sharing; enhanced negotiating power by combining purchases; greater opportunities for superior product and service creations via interdisciplinary teams, all leading to business regeneration and growth. These advantages, derived from the *collaborative power of teams,* can be an enormous differentiator in making your organization a recognized leader in your industry or service sector, as opposed to being a mediocre also-ran.

Customer Driven

Customer driven means demonstrating a superior ability to understand, attract, and keep valuable customers. It also means having the business agility to change and evolve along with your customers, responding to shifting needs, market changes, and new opportunities as they arise.

Xerox Corporation's Quality Policy, shown in the sidebar, articulates the fusion between quality and a focus on the customer.

> Xerox is a quality company. Quality is the basic business principle for Xerox. Quality means providing our external and internal customers with innovative products and services that fully satisfy their customer requirements. Quality is the job of every Xerox employee.

This Quality Policy is like the North Star. It gives all Xerox employees a vision to relate to, and it guides behavior. It also helps break down barriers by setting expectations that people from different functions need to collaborate to meet customer requirements.

Quality function deployment is a rigorous process used by progressive companies to help internal work teams increase their focus on the external customer. The customer's voice is used to determine what the critical properties of a new product or service should be. This method leaves nothing to the imagination in determining what will satisfy customers, how the new product or service stacks up against the competition, and how internal work processes must be managed to guarantee meeting customer needs.

WHAT MAKES THE TRANSFORMED ENTERPRISE TICK? COLLABORATION!

If you go back and reread the information on the five characteristics defining what I've called the "transformed enterprise," you'll discover not only how tightly interwoven they are, but also that the theme of *collaboration* is associated with each one. Collaboration then is the glue that holds the whole scheme together and makes it work.

Ken Wright underlines this point in his book *The People Pill.*

> Collaboration is particularly crucial to today's business environment. As globalization and competition intensify, business becomes more about relationships and networks, making internal and external collaboration essential. There will be more time spent applying critical thinking and really understanding the effects of collaboration, and in this new world people will need to develop personal leadership skills to help improve the performance of increasingly scattered work groups and to maximize the effectiveness of outsourcing as globalization and decentralization take hold. Without an effective collaboration plan, one hand will not know what the other is doing.... Successful businesses of the future will overcome these issues by linking all areas of the business together to insure a collaborated approach.[3]

In summation, regardless of the product or service you sell, your customers have changed. Their demands are lengthening; their patience is shrinking. Huge shifts in the global economy have given them increased power to command exactly what they want, the way they want it, when they want it, at a price that will make you cringe. You'll either meet these demands or be out of business.

Your organization's success in gaining and holding a competitive advantage so you can stay in business in a complex, global, whitewater environment lie as a *gold mine of ideas* in the heads of all your people. Being an ongoing competitive force means not just keeping up with the pace of change—but rather, capitalizing on it to better satisfy your customers' multifaceted requirements.

To accomplish this, your organization must leverage the synergy and collaborative brain power of your employees continuously, in many configurations, at all levels, within and across all functions. The days of viewing employees simply as "pegs" to be slotted here or there and spoken of as impersonal "headcount" are well behind us. It is through collaboration that human ingenuity and creativity are best utilized. Today's problems are too enormous, the pat answers too few, and the stakes too high for it to be any other way. And tomorrow's problems will only raise the ante for organizational collaboration.

A final thought before moving on. Although the Boston Celtics have won 17 world championships, including 8 in a row from 1959 to 1966, they have never had the league's leading scorer and never paid a player based on his individual statistics. The Celtics understand that virtually every aspect of winning basketball requires close collaboration among the players, coach, and front office.

NOTES

1. S. H. Simmons, *How to be the Life of the Podium* (New York: AMACOM, 1982), 249-250. According to Ms. Simmons, "Leo Rosten, writing in the *Saturday Review,* originally recounted this story about an unknown wit analyzing the operations of a symphony orchestra for technical efficiency."
2. S. F. Dichter, "The Organization of the '90s," *The McKinsey Quarterly,* no. 1 (1991), 146–147.
3. K. Wright, *The People Pill: Proven to Cure the Headache and Heartache of Engaging People* (Chatswood, New South Wales, Australia: Amanda Gore International, 2009), 38–39.

CHAPTER TWO

THE CORE ELEMENTS FOR COLLABORATIVE PARTNERSHIPS

Six Ingredients Required for Success

CHAPTER OBJECTIVES

- To define collaboration and present an example to demonstrate its power
- To present a model setting forth the six elements that must be accounted for in creating collaborative partnerships, and to detail each one with examples
- To argue the case for collaboration and present cautions when implementing it

INTRODUCTION

Pat Riley, a five-time championship head coach in the National Basketball Association and now president of the Miami Heat, once said something about hard work and practice that also rings true for collaboration: "While hard work and practice won't guarantee you anything, without it, you can't even begin to think about successfully competing in the NBA."

So it is with collaboration. While collaboration won't guarantee anything, without initiating, maintaining, and refining it throughout your organization, you can't even begin to think about successfully competing in the whitewater business and economic environment we discussed in the previous chapter. And if collaboration is to be the name of your game in order to become, or remain, world class in your marketplace, you need collaborative leaders who know how to build collaborative partnerships.

To have collaborative leaders means you and other executives and managers need to embrace a fundamental belief that no single person—or elite senior-level staff group—can possibly have all the answers on how to capitalize on every key opportunity in an ambiguous, competitive marketplace. Then, based on this shared belief, you and the other executives and managers need to actively promote collaboration at all levels, in every nook and cranny, to innovate and develop new solutions that improve business results.

If you're in a market situation that allows you to operate with a more traditional bureaucracy, *collaboration will make you better.* If you're now in the more common situation of a dynamic, global marketplace trying to make a transformed enterprise—one that is flattened, more flexible, fast acting, team oriented, and customer driven—function at its full potential, *collaboration is an absolute necessity and collaborative leaders are the only ones who can make it happen.*

LESSONS FROM GEESE

The example of geese is often held up as collaboration in its most pure and genuine form. I first saw this narrative on a 1986 United Way campaign poster,[1] at which time I copied it down. I have used it countless times since in seminars and workshops. As you read this story of geese in the sidebar, you will appreciate the power of collaboration as never before.

As I repeatedly tell managers in my workshops, building collaboration into your organization is not too difficult. All you need are a bunch of people who have the same values as a gaggle of geese: working in harmony to maximize the output of the team,

The next time you see geese heading south for the winter, flying along in V formation, you might be interested to know what science has discovered about why they fly that way. Researchers have learned that as each bird flaps its wings, it creates uplift for the bird immediately following. By flying in a V formation, the whole flock adds at least 71 percent greater flying range than if each bird flew on its own.

Whenever a goose falls out of formation, it suddenly feels the drag and resistance of trying to go it alone and quickly gets back into formation to take advantage of the lifting power of the bird in front. When the lead goose gets tired, it rotates back in the wing and another goose flies point. The geese honk from behind to encourage those up front to keep up their speed.

Finally, when a goose gets sick, or is wounded by gunshot and goes to the ground, two other geese fall out of formation and follow it down for protection. They stay until it is able to fly or is dead, and then launch out on their own or with another formation to catch up with their original group.

respecting the personal drag of trying to go it alone, encouraging each other, rotating leadership as conditions change, and sacrificing personally to provide help unconditionally to team members experiencing difficulty!

FRAMING THE ESSENCE OF COLLABORATION

The literal, sterile definition of collaboration is "to co-labor; to labor together." But that definition is too simple for our purposes. Over time, through many debates and discussions, I've evolved the following definition, which has proven to be quite practical in my work with teams and organizations.

Collaboration is a joint effort between two or more people, free from hidden agendas, to produce an output in response to a common goal or shared priority. Often this output is greater than what any of the individuals could have produced working alone.

Collaboration works best when relationships are treated as genuine partnerships. That is, all parties are involved in creating superior new value together rather than merely performing one-for-one exchanges where a person simply gets something back for what is put in. In a true collaborative partnership, obligations are broadly distributed, the possibilities for cooperation are more extensive, understanding and solidarity grow among the collaborative partners, communication is frequent and intensive, and the interpersonal context is rich.

An executive once gave me this interesting perspective on collaboration. He said, "For me, the bottom line measure of collaboration is the ratio of *we's* to *I's* that I hear in my interactions with people throughout the various departments. If I get the sense I'm hearing more *we's* than *I's,* I have at least one indication that collaboration is alive and well in my division." A true collaborator thinks and practices *we* before *I.*

SIX INGREDIENTS REQUIRED FOR COLLABORATIVE PARTNERSHIPS IN ORGANIZATIONS

During my career at Xerox, I conducted hundreds and hundreds of team building sessions involving several thousand managers from executives down to first-line managers. While the problems and the desired outcomes for each session were different, often the issue of intra- or interteam collaboration came up for processing.

When that was the case, I'd begin the team's self-discovery and analysis by first reviewing the definition of collaboration shown previously. Then I'd ask, "Given our definition, what ingredients are essential for collaboration to grow and flourish within or across teams?" As participants shared their viewpoints round-robin style, I wrote them on flip charts, and then we combined them into themes. While just the first step in working the problem, it oriented the team to a bigger picture: Collaboration doesn't just happen; "the bed has to be cultivated before the flowers can grow."

Over time some clear-cut themes began to emerge; these were then shared with successive teams to consider and to process along with their own perspectives. Eventually, six solid collaborative ingredients evolved that have stood the test of time in my work. The conceptual model depicted in Figure 2-1 highlights and integrates the six ingredients necessary for building collaborative partnerships inside work teams, within interdisciplinary project teams, as well as between people and teams in different units. This model demonstrates how the "bed is made" for collaboration to flourish.

Figure 2-1. Six Ingredients for Collaborative Partnerships in Organizations

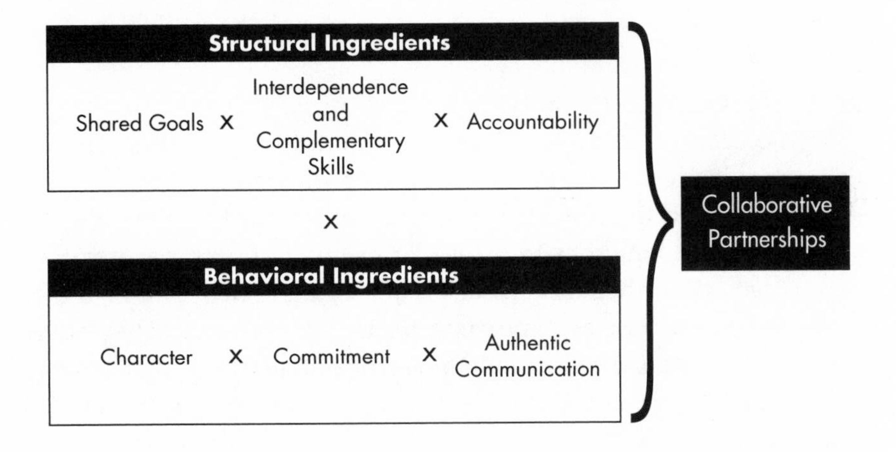

I've broken the six ingredients into two natural categories to make the model easier to understand. One set of three dimensions form the "Structural Ingredients," and the other group of three dimensions comprise the "Behavioral Ingredients." The Structural Ingredients are the mechanics of collaboration and need to be planned and orchestrated to ensure that they are in place. The Behavioral Ingredients represent the relationship values all people need to bring to the task for successful collaboration to occur.

Notice how the ingredients within and across both categories are shown as being multiplicative, not additive. This signifies the synergistic effect of collaborative power. Remove any one of the dimensions, and the multiplier effect is drastically reduced. All six ingredients bolster each other to produce high-powered, collaborative partnerships.

Let's examine each of ingredients in more detail and demonstrate why all six are fundamental to successful collaborative partnerships in organizations.

THE THREE STRUCTURAL INGREDIENTS

We will begin with the three structural ingredients and then move on to the three behavorial ingredients.

Shared Goals

For collaboration to take root, any project's goals (or goal) must be seen as important by the collaborative partners. Therefore, as a leader trying to set up a collaborative partnership, you not only need to be sure shared goals exist, but that they align with some personal desire or work objective of each partner. Shared goals not only define what we are trying to do together, they also create purpose and meaning for the larger cause. They energize the partners intellectually and emotionally to buy into the principle: "This team's success is my success and my success is this team's success."

Morten Hansen in his book *Collaboration* writes about the role of shared goals in terms of unifying people in a collaborative effort.

> Leaders who practice disciplined collaboration translate their collaborative aspirations into a unifying goal.
>
> Leaders must craft a compelling unifying goal that makes people commit to a cause greater than their own

individual goals. A crafty common goal meets four tests: it must create a common fate; it must be simple and concrete; it must stir passion; and it must place competition on the outside.[2]

Using common language, symbols, storytelling, and metaphors, you can bring clarity and understanding to the goals and evoke positive emotion for them. The articulated shared goals become a rallying cry for all partners.

Interdependence and Complementary Skills

These two go hand in hand. Without them actively in place, collaboration grinds to a halt.

Independent big egos coupled with redundant skills leads to either unresolved conflict or stalemate as people, worried about their roles in the endeavor, fight to maintain their position and self-esteem. As many organizations know only too well from experience, this certainly is not a formula for success in serving customers, clients, or the citizenry.

Regarding *interdependence*, collaborative partners require each other's knowledge, skills, abilities, and experiences to achieve the task or project's shared goals. Shared goals mean no teammate can accomplish alone what the partnership can accomplish together.

Likewise, *complementary contributions* are a necessity to achieving the shared goals. Each collaborative partner is considered an asset because each brings something different of value to contribute to the relationship. Their motives for entering the relationship are positive—to pursue an opportunity that will improve the larger business—not negative: to pursue a selfish goal at the expense of the larger business, or pursuing a goal that undermines the larger business.

Interdependence and unique complementary skills are vital to successful collaboration. As Jim Collins noted, "Get the right people on the bus, the wrong people off the bus, and the right people in the right seats."[3] Curtis Carlson and William Wilmot, in their book *Innovation*, elaborate even further.

> Only interdependent team members with unique complementary skills and who can collaborate have a place on a high-performance innovation team. Every member must feel secure about having a significant role in the project. After all, you can't dance with someone stepping on your toes. Ambiguity about the role of each team member prevents the commitment and collaboration for team success.
>
> Innovation teams are powered by the collective intelligence of the members of the team. When team members are engaged in an iterative value-creation process, they can achieve customer value tens, hundreds, or thousands of times greater than any individual could achieve alone. But each team member must bring critical skills that are necessary for the success of the project. Each individual on the project must clearly understand his or her own role and importance to the success of the project.[4]

Two examples of cross-functional collaboration from Proctor and Gamble demonstrate the powerful synergy of interdependence and complementary skills. P&G melded the efforts of specialists from different units to initiate improved customer value and enhanced business performance by innovating Crest Whitestrips and Olay Daily Facials.

Developing Crest 3D Whitestrips involved collaboration among individuals from the oral-care area who were experts in teeth whitening, people from the fabric and home-care area who were experts in bleach properties, and scientists from corporate research and development who were able to create some novel film technology.

With Olay Daily Facials, P&G wanted a product that provided both excellent cleansing and moisturizing. Again, people from multiple internal units were brought together. Skin-care specialists who understood the surfactants needed in facial cleansing joined with professionals from the tissue and towel area who brought in substrate knowledge. And people from the fabric-enhancer area involved with Bounce were key contributors because Bounce used a similar technology for putting fragrance on clothes.[5]

Accountability

No group of individuals can ever become a team of collaborative partners until they unequivocally embrace the dimension of accountability for the results they collectively produce. This is a demanding test because accountability serves as an invisible structural link among every partner. Accountability binds them as one, advances commitment, puts a premium on individual trustworthiness, and slams the door on future backpedaling or CYA (cover your ass) excuse making should things get sticky or failure be a possibility.

Accountability in a collaborative partnership arises from the time and energy teammates spend in vigorous debate and discussion devoted to understanding and taking ownership of the shared goals and in figuring out how best to accomplish them together.

On the other hand, people in a *compliant group* don't have the opportunity to do that; they do what they are told to do. As such, the individuals feel little to no accountability for the results they produce; so the boss has to push each person hard to put out enough effort to get a passable job completed. At its very best the final output will only equal the sum total of the individual inputs. More often than not even this paltry output is not achieved. The typical end result turns out to be an output well below the combined additive potential that each individual contributor could have put forth.

In a collaborative team, the partners hold themselves accountable for maximizing their individual contributions *and* for multiplying those individual contributions into a synergistic output greater than could have been produced working as an additive, compliant group.

The leader of a collaborative team does not have to drive and push for results. Instead, the collaborative leader makes sure the team is perking along at the highest level by acting as a facilitator, helper, barrier buster, and cheerleader. Accountability rests with all the teammates, *and the leader is considered a vital teammate with important complementary knowledge and skills*. In collaborative partnerships, "we all are in this together; we all hold ourselves accountable, as a team, for our results."

In her book *Fearless Leadership,* Loretta Malandro emphasizes the need for accountability in committed partnerships.

> Committed partners accept 100% accountability on all matters. This means keeping your attention on what you can change, not what *others* should do.
>
> ... [With 100 percent accountability] it does not matter what others choose. Accountability is viewed as owning the problem or situation. It is irrelevant who is to blame or what caused the problem. With the "owner mind-set" firmly in place, people quickly resolve problems by confronting breakdowns and blind spots and operating with an enterprise perspective.[6]

A leader and team members who fully commit to each other and assume full accountability for their actions and results develop linkages and shared ways of operating so they can work together smoothly. They build broad connections between many teammates at many organizational levels and become both teachers and learners, all with the intent of strengthening themselves to be even stronger partners so they can assume more accountability in the future.

THE THREE BEHAVIORAL INGREDIENTS

Next let's look at the three Behavioral Ingredients—character, commitment, and authentic communications—that are also fundamental to successful collaborative partnerships in organizations.

Character

Character, or integrity, is the most critical of the behavioral elements. And if you think about it, how could collaborative partnerships ever be formed and operate in a reliable manner without all people involved being of high character? Short answer, they can't. And the reason is, character is a key to building and sustaining mutual trust; and mutual trust, in turn, is *the behavioral core* of collaboration with no strings attached.

Noel Tichy and Warren Bennis make clear what is meant by having character (integrity).

> [It] means having values. It means having a moral compass that sets clear parameters for what one will, will not do. Character is all about knowing right from wrong and having worked these issues out long before making tough judgment calls. It's about knowing what your goals and standards are and sticking with them.
>
> For us character also means putting the greater good of the organization, or society, ahead of self-interest. As Peter Drucker put it, it is worrying about "what is right" rather that "who is right."
>
> Character is that distinctive, unfiltered personal voice that cannot be faked or imitated. It is the core of who we are.[7]

The late UCLA basketball coach John Wooden always urged, "Be more concerned with finding the right way than having it your way." Yes, character is front and center relative to collaboration, and it's more than just mouthing the right words. It means acting toward each other in honorable ways that justify and enhance teamwide mutual trust, it means holding deep-seated values about the power of collaboration that are above reproach, and it means living those ideals by "checking your ego and selfish interests at the door" to be an essential part of a relationship striving together to expand 10 + 10 from equaling 20 to 10 x 10 equaling 100 or more!

Commitment

Commitment builds from each person being counted on to be an active partner in the experience of creating something of value together. Commitment means teammates are willing to obligate themselves to a truly engaging purpose, larger than just personal self-interest. They are willing to give of themselves to achieve a common project goal or to help the collective enterprise succeed. When that internal fire burns, partners easily demonstrate their commitment to the collaborative relationship by investing in each other in pursuit of *their* shared goals. They invest in three ways—often at great personal sacrifice—for the greater good.

Partners *invest tangibly* by devoting time, money, people, facilities, and other resources to the relationship. They *invest intellectually* by freely sharing knowledge and information at their disposal, as well as their ideas and their considered perspectives. They *invest emotionally* by caring for and about each other as human beings, by being both teachers and leaders, by "having each other's backs," by being trustworthy.

The collaborative energy that flows from commitment based on the three levels of mutual investment—tangible, intellectual, and emotional—is unbeatable and is the fuel for success.

Price Pritchett, in his book *Firing-Up Commitment during Organizational Change*, paints a vivid word picture of the significant role commitment plays in any collaborative endeavor.

> Commitment energizes. It empowers. It inspires creativity and pulls a person's true potential into play.... Commitment is self-nourishing. Even as it draws power from the spirit it feeds the soul. Commitment gives meaning to work, and deepens one's sense of self-worth.
>
> [And regarding the leader's role in generating commitment,] commitment climbs when people see passion in the person out front. Your intensity—your focus, drive, and dedication—carries maximum influence over the commitment you can expect from others.... If you provide lukewarm leadership, you'll see the passion cool among your people. Commitment can't survive when the leader doesn't seem to care. So be obvious. Turn up the burner inside yourself. Let the heat of your commitment be strong enough to glow in the dark.[8]

Authentic Communications

Authentic communications are open, honest, nonjudgmental, and nonevaluative; they attack issues, not people. Collaborative partners understand "messengers who bring bad news are *not* shot" because "shooting" them stifles the free-flowing communications that must be cultivated in any ongoing collaborative relationship. Collaborative partners routinely propose and build on ideas; they transparently seek and share information; they listen closely to each other; they test for understanding; they use constructive disagreement and fruitful friction to avoid groupthink, conformity, and compromise.

Authentic communications has to be accepted as a non-negotiable rule within collaborative partnerships because this is the surest way: to show respect for each other, to prove no single person thinks he or she has all the answers, and to minimize the promotion of self-interests.

The late Bill Walsh, who guided the San Francisco 49ers to 3 Super Bowl championships and 6 NFC West division titles in his 10 years as head coach of the 49ers, makes a number of compelling comments regarding communications and collaboration in his excellent book *The Score Takes Care of Itself*.

> Quality collaboration is only possible in the presence of communication; that is, the free-flowing robust exchange of information, ideas, and opinions.
>
> For me that meant I had to set aside certain aspects of my ego—e.g., talking too much—and really listen to what talented individuals in the organization had to say. I had to learn that communication is not a one-way street; it's a two-way, three-way, every-way street.
>
> I was never called Coach Walsh. In fact, everyone in the organization was addressed by their first name, including me. I wanted no barriers such as rank or title to clog up the productive interaction, no chain-of-command to produce a sense that instead of a real team we were just a collection of isolated individuals on a totem pole of power belonging to small independent units. Rank, title, or inferred status can impede open communication in an environment where people thrive on helping one another.
>
> When you reach the point where someone in your organization comes up with an idea better than the one you

have been extolling for weeks or months, and it makes you happy, you're an authentic communicator and collaborator.[9]

From a *technical side,* authentic communications means vigorous, full-bodied debate and discussion around: functional strategies, business goals, technical information, market data, financial data, trends, improvement ideas, and the like. From a *human side,* this includes sharing knowledge and processing information and ideas concerning: personal conflicts, performance issues, trouble spots, workload, changing situations, promotional opportunities, and other similar subjects.

The Six Ingredients Model is highly instructive because it clarifies the underpinnings of collaborative partnerships. It shows all the ingredients as interlaced, yielding a considerable multiplier effect that produces the synergy inherent in genuine collaborative partnerships. The discussion following the conceptual model itself delineated the characteristics of each dimension and how it operates in practice.

POSITIVE CONSEQUENCES OF ORGANIZATIONAL COLLABORATION

The cornerstone of building a collaborative culture is teaching your people how to make collaboration happen—how to facilitate it. That's what the remaining chapters of this book will teach you. Also, collaboration cannot be practiced by only one segment of your organization—it has to be everywhere. It has to flow up, down, sideways, and diagonally, both inside and across every function in order to obliterate the internally competitive and destructive win/lose, attack/defend, us/them conflicts that are standard operating procedure inside so many of today's organizations.

Your enemies are not within your organizational boundaries. They are outside your boundaries. Your enemies are your external competitors, and they are formidable. They are after your people and your market share. Spending any time and energy erecting walls and defending turf in noncollaboration, political gamesmanship, and destructive conflict, is deadly. It will bring your company to its knees.

Tim Brown, CEO of IDEO, and a leading management thinker, has some provocative thoughts on this subject of organizational collaboration and innovation.

> The lone designer, sitting alone in a studio and meditating upon the relationship between form and function, has yielded to the interdisciplinary team.
>
> [It] is common now to see designers working with psychologists and ethnographers, engineers and scientists, marketing and business experts, writers and filmmakers. All of these disciplines and many more have long contributed to the development of new products and services, but today we are bringing them together within the same team, in the same space, and using the same processes. As MBAs learn to talk to MFAs and Ph.D.s across their interdisciplinary divides (not to mention the occasional CEO, CFO, and CTO), there will be increasing overlap in activities and responsibilities.[10]

Like a stone cast into the middle of a pond that causes radiating ripples to occur, collaboration cultivated and promoted in the middle of an organization can bring about its own radiating ripples of influence and power. Barriers, real and perceived, complex and simple, tumble down under the onslaught of collective thinking; the white space between the boxes on the organization chart begins to diminish; the frozen chain of command, where the top dictates to the middle what to force on the bottom, begins to thaw and melt; and informal relationships—consisting of alliances and networks among people that actually get the work done—are recognized and fostered by providing needed resources to help them flourish.

Moreover, high payoff opportunities are uncovered and advantage is taken of them. Joint marketing between divisions, a product variant linking components from units that traditionally sold their wares separately, or a procurement staff for one business unit helping another find a supply source for a new venture, are just a few examples.

CAUTIONS REGARDING ORGANIZATIONAL COLLABORATION

I don't want to give the impression that collaboration is a cure for every organization ill and all you to have to do is brandish your "magic conductor's baton" and all your employees will work in harmony forever. There are important cautions to be mindful of as you think about collaboration and move forward in your effort to be a collaborative leader. So let's cross to the other side of the street and examine the *cautions regarding collaboration*.

The first caution: do not slough off the six elements discussed earlier. You need to manage those elements and put a mental checkmark by each one after giving it your full consideration when planning and organizing any collaborative project.

The second caution: do not overuse collaboration. Many tasks are better done by an individual or better led by an experienced person who directs and oversees others in carrying out the task according to the expert's specifications.

The third caution: if after a project gets underway, and you have "applied some grease to a squeaky wheel" and it keeps squeaking, replace it. You can't let one or two people, for whatever the reason, destroy the success of a collaborative team working to produce something of value on behalf of the organization.

The fourth caution: recognize the current culture you work in. Some cultures will be ripe and ready for gung ho collaboration; other cultures will be ready to proceed but will be more conservative and be a bit "old school," requiring a more careful and measured approach; and finally, there will be the macho, individualistic cultures, or ones run by an autocratic patriarch or matriarch, that will make it nearly impossible to do much collaborating. Know your culture and proceed accordingly.

Further cautionary insight about collaboration can be found in the transcript of a podcast interview by Peter Shaplen for News @ Cisco held with Morten Hansen, professor of management at UC Berkeley and author of the book *Collaboration*. The interview—which I have excerpted and added headers to highlight just three key points—was entitled "Business Collaboration: Getting It Right."

Right Way to Collaborate vs. Wrong Way

Morten Hansen: There's a right way to collaborate, and there's a wrong way to collaborate, and what leaders have to do is to instill the right way and avoid the traps of collaboration, and that's a very, very different approach than simply saying, "We need more of it."

Peter Shaplen: How does one make a distinction between good collaboration and bad?

Morten Hansen: Bad collaboration is about collaborating without a strong focus on results, so you're collaborating for the sake of collaborating. Bad collaboration is about launching a collaboration project, hoping it goes well, when, in fact, it's going to lead to infighting and fighting over agendas and objectives without focusing on the work, and in those cases, you're oftentimes better off not collaborating but doing the work yourself.

Peter Shaplen: How do you determine what is the right way and the right result?

Morten Hansen: The goal of collaboration is not collaboration itself. It is better results. We are after results.

Don't Collaborate When It Doesn't Make Sense

Peter Shaplen: Are there times when . . . it's okay not to collaborate?

Morten Hansen: Absolutely. . . . The executive first has to ask the question, "What is the upside for us?" And usually the upside is in three buckets: First is the innovation upside. You can work better across the organization and innovate better. Second is the customer upside, particularly the cross-selling. You can sell more products to your customers by coordinating better. Third is the efficiency upside. Efficiency is gained by best practice transfer and not reinventing the wheel all the time in the company.

So you go through an exercise and see where the upside is in our company. Is it big enough and in which of these areas? Sometimes you come to a conclusion that it isn't big enough, and then you should have the discipline to say, "We should not collaborate. It is not our key priority right now."

Choose Appropriate People

Morten Hansen: So the first thing when you start a collaboration project is to ask the question, "Who can contribute?" The great thing about collaboration is that some of these people might be junior people sitting in very different locations, and you have an opportunity to bring those people together. Collaboration provides you with that power, and what you want, of course ... is to take people who have the complementary expertise and bring them together. That increases performance. But very often, the hierarchy of a company gets in the way.

Disciplined collaboration is about focusing on the result and asking the question, "Who can contribute here?" and usually it is a selective group of those who have the expertise. It is not everybody, and it's okay to say, "Well, these 10 people are perfect for this project, but these other 20, who may want to join, are not the best for this project."

We don't want 30 people in a room just because everybody wants to participate.

COLLABORATION DOESN'T JUST HAPPEN, IT REQUIRES PLANNING AND FACILITATION

Collaboration is not a pipe dream; it is real; it is a difference maker in terms of business effectiveness. However, extolling its virtues and urging executives, managers, professionals, and individual contributors to be more collaborative won't cut it. It takes planning and facilitative leadership to bring everything together and make collaboration operate at its full potential within an ongoing work team, an interdisciplinary project team, or across functions as executives attempt to resolve an ad hoc problem that is sapping the synergy between their units.

The "bed" for superior collaboration must be prepared first. That means anyone assigned leadership for a cross-disciplinary project or desiring to upgrade task collaboration within her work team has some substantial work to do. The leader first must contemplate and design the structural elements of collaborative partnerships: shared goals, interdependence and complementary skills, and accountability to be certain they reinforce each other and will act as the engine driving the collaborative effort.

Then within the planned structure the leader must think long and hard about what Morten Hansen said: "Who can contribute here?" Bringing people onboard who fulfill the behavioral side by having the necessary character, commitment to the shared goals, and the authentic communication skills to be worthy collaborative partners is crucial. While many may be considered, the fewest number of those "with the right stuff" needed to achieve the shared goals should be chosen.

With the collaborative bed fully cultivated, the leader then has to move into a leadership role as a facilitator, not a commander and controller. The solutions necessary to achieving the shared goals lie within the team members. Therefore, the facilitative leader's task is to promote, unleash, and leverage the wisdom, innovation, and creativity each teammate brings to the collaborative partnership.

Mining gold nuggets of wisdom in a collaborative manner takes some learning and know-how. It may require behavioral shifts for you and a number of your people, especially your managers. One thing is clear to me, however: facilitating collaboration to gain a competitive edge for your organization must be a shared responsibility among as many different executives, managers, professionals, and individual contributors as possible.

The next chapter ties together the ideas from the first two chapters. It also sets the stage for the rest of the book by providing an integrative framework that links the requisite collaborative leadership skills I've found essential to facilitate collaboration successfully within single work teams, interdisciplinary projects, and in cross-functional joint efforts.

NOTES

1. The reference on the United Way campaign poster read as follows: "Adapted from Barbara Stirling Willson."
2. M. T. Hansen, *Collaboration: How Leaders Avoid the Traps, Create Unity, and Reap Big Results* (Boston: HBR Press, 2009), 91.
3. J. Collins, *Good to Great: Why Some Companies Make the Leap and Others Don't* (New York: HarperCollins, 2001).
4. C. R. Carlson and W. W. Wilmot, *Innovation: The Five Disciplines for Creating What Customers Want* (New York: Crown Business, 2006), 191–192.
5. "At PG It is 360-Degree Innovation," *BusinessWeek*, October 11, 2004.
6. L. Malandro, *Fearless Leadership: How to Overcome Behavioral Blind Spots and Transform Your Organization* (New York: The McGraw-Hill Companies, 2009), 149–150.
7. N. M. Tichy and W. G. Bennis, *Judgment: How Winning Leaders Make Great Calls* (New York: Penguin Group, 2007), 70–71.
8. P. Pritchett, *Firing-Up Commitment during Organizational Change* (Dallas, Texas: Pritchett & Associates, Inc., 1994), 2, 4.
9. B. Walsh with S. Jamison and C. Walsh, *The Score Takes Care of Itself: My Philosophy of Leadership* (New York: Penguin Group, 2009), 112, 115, 116.
10. T. Brown, *Change by Design: How Design Thinking Transforms Organizations and Inspires Innovation* (New York: Harper Business, 2009), 26.

CHAPTER THREE

AN INTEGRATIVE FRAMEWORK LINKING THE BOOK'S CHAPTERS

Your Pathway to Unleashing the Collaborative Genius of Work Teams

CHAPTER OBJECTIVES

- To provide a conceptual illustration of an integrative framework that links the book's chapters and sets the context for understanding collaborative partnerships
- To couple the illustration with an explanation of the pieces and their linkages

INTRODUCTION

Chapter Three is a bridging chapter. It provides an integrative framework that helps associate the content of the first two chapters with the content that is to follow. The illustration shown in Figure 3-1, on page 39, depicts the three segments comprising the book's integrative framework along with corresponding chapter numbers as reference points. Figure 3-1 is a road map for keeping the learning content of *Building Team Power* organized as you continue to read.

QUICK REVIEW: THE TRANSFORMED ENTERPRISE

The top left of Figure 3-1 on page 39 shows the five linked elements of the transformed enterprise from Chapter One. Evidence was presented that a whitewater global economy has increased marketplace uncertainty, complexity, and interdependence for most organizations today. It's true, in limited instances—depending on the sector and the goods or services produced—that some firms may face relatively stable, high-volume, mass market conditions. In those situations the traditional bureaucracy can be appropriate and work for them.

In contrast, organizations constantly buffeted by shifting global markets, ever-changing customer requirements, rapid technological change, quick product obsolescence, falling trade barriers, ferocious competition, and the like, have discovered the traditional command-and-control bureaucracy is a poor fit with these realities. These dynamic pressures overload the bureaucracy's ability to respond.

Organizations facing the whitewater conditions of today are shifting away from the traditional hierarchy to a new way of doing business. Although bureaucratic vestiges remain at the core, a *transformed enterprise* has emerged that is being widely embraced. It is flatter, more flexible, faster, team oriented, and customer driven. The transformed enterprise is much more adaptable, better able to tap the innovative power of its people and technologies, and when coupled with a collaborative culture, this change can become a competitive advantage.

QUICK REVIEW: THE SIX INGREDIENTS FOR COLLABORATIVE PARTNERSHIPS

The top right section of Figure 3-1 portrays the six foundational ingredients discussed in Chapter Two for building and nurturing any collaborative effort. Whether within work teams, within interdisciplinary project teams, or across functions or divisions, successful collaborative partnerships require the three structural dimensions (shared goals, interdependence and complementary

Figure 3-1. Integrative Framework for Building Team Power

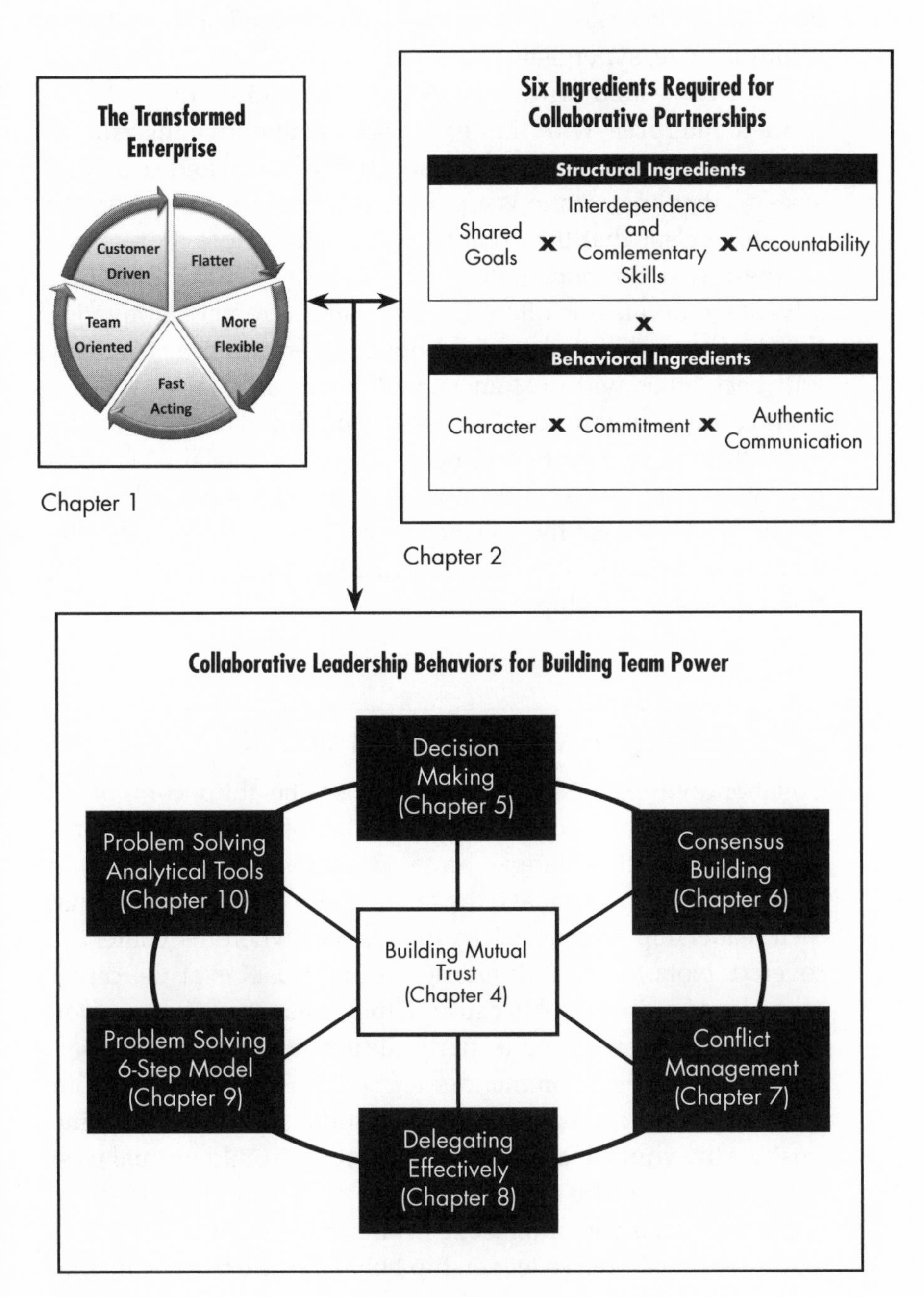

skills, and accountability) and the three behavioral dimensions (character, commitment, and authentic communications) all to be present. These six ingredients are intertwined and produce a multiplicative, synergistic result.

The two-headed arrow connecting the diagram of the transformed enterprise with that of the six collaborative ingredients indicates a highly integrated architecture of social and structural systems that organizations are obliged to set in place today. It is becoming clear that the organizational architecture represented by these two components can be a key source of competitive advantage because it offers both the structure and motivation that enables individuals and groups to interact more effectively with each other, with customers, and in teams.

As companies gain equal access to capital, and many technologies continue to mature and become generic, sustainable competitive advantage depends more and more on the capability to deploy and leverage the collaborative efforts of the people who make up the organization. And that leads to the third piece of the framework—leadership.

COLLABORATIVE LEADERSHIP BEHAVIORS

Collaborative leadership skills comprise the third component of our integrative model. The bottom half of Figure 3-1 pictures seven connected rectangle boxes designating the leadership behaviors required to make the architecture operate. Notice that each leadership rectangle lists the chapter where its content is covered. Note too that "building mutual trust" is at the center of the leadership model because without that as your basis, you *cannot* be a collaborative leader building collaborative partnerships. You can be a commander and controller relying on your position power to get compliance from others, but without being trustworthy yourself and having the ability to build mutual trust with others, you'll be incapable of being a collaborative leader.

The vertical arrow intersecting the horizontal arrow specifies that these collaborative leadership behaviors are the engine that drives collaboration across the organizational architecture. In other words, we can make changes to the reporting relationships making up the formal structure; we can alter lines of authority; we can spell out new roles; and we can modify work policies to conform to the transformed enterprise; in addition, we can exhort our people to work together and be mindful of the six collaborative ingredients.

However, while these are essential steps in the total scheme of things, they are not enough. For collaborative partnerships to flourish as a way of life, collaborative leadership is necessary. Units need to be filled with leaders who both role-model the six basic ingredients as they plan and organize collaborative partnerships, and who champion, orchestrate, and facilitate collaboration using the seven leadership behaviors.

There are numerous problems individuals and teams encounter as they attempt to work together. Problems and issues like these lurk everywhere: (1) confusion over, or disagreements about, what the goal of the team really is; (2) "hidden agendas," individual goals, ambitions, and feelings that are not shared with the group yet influence the person's behavior; (3) resentment about "giving up" what teammates perceive to be their territory; (4) unwillingness to accept one's defined role in the team; (5) disagreements over procedures—how best to complete a task; (6) strong competitive feelings among some teammates that significantly hinder their collaboration; and (7) a climate in which teammates are afraid to voice their feelings, opinions, and ideas.

These are just a sample of issues underlining the main point: building and maintaining collaborative partnerships is not "DNA instinctive" in organizations. You need a structure and work processes that set the context for it, you need to understand and set in place the six ingredients as the foundation for any collaborative relationship, and you have to lead it; coach it, and facilitate it to make it all work.

For that reason, the seven leadership behaviors for creating and sustaining collaborative partnerships between people and teams constitute the remaining chapters of this book. I have come to believe that building mutual trust, making decisions, building consensus, managing conflict, practicing effective delegation, and utilizing team problem solving are the core skills of collaborative leadership. And if understood and practiced authentically, they will give you the leadership know-how to be a collaborative leader in your organization. Each chapter's content is aimed at providing simple models, tools, and processes, but always with emphasis on *doing it collaboratively*.

There is an old story about three men fishing on a small lake. The owner of the boat watches in astonishment as his two companions, one after the other, climb out of the motorboat and walk across the water to a lakeside bar for a few beers. Assuring himself that his faith is as great as theirs, the owner steps out of his boat and promptly sinks. The first fisherman turns to the second and says, "Do you suppose we should have shown him where the rocks were?"

The sidebar story provides a fitting lead-in to the upcoming skill-building chapters.

Like our little tale, facilitating collaborative partnerships is not a simple matter of getting out of an old motorboat, strolling across the water into a team, giving high-fives, drinking a few beers together, and then sitting back to watch a beautiful collaborative partnership unfold and flourish. You need to know where the rocks are.

That's what the remaining chapters will cover. I'll show you where the rocks are so you can use them as stepping-stones for *building team power through collaborative partnerships for innovation and results.*

Part II

Seven "How-to" Collaborative Leadership Behaviors for Building Team Power

CHAPTER FOUR

MUTUAL TRUST

The Heart of Collaborative Leadership

CHAPTER OBJECTIVES

- To present the 3-Cs of trustworthiness: competence, character, and communications
- To showcase a proven model for building mutual trust and detail its elements
- To review the major trust breakers and their negative consequences on relationships
- To provide compelling evidence why trust building is the core of collaborative leadership

INTRODUCTION

During the summer between my junior and senior years in college, I worked on the shipping and receiving dock of a small, family-owned manufacturing firm that employed about 250 people. The president and owner of this company—we called him "King Ironfist"—trusted no one but himself to do anything.

King Ironfist micromanaged everyone in every operation, he dominated and cut off the sharing of important data by other people, he was the consummate "hip shooter," he was full of the right answers, he never asked for advice and rejected all that anyone dared to offer, he made many promises to employees that were quickly broken or soon forgotten, and when he made mistakes, he tried to cover them up or blame someone else. Except for a few cronies, no one trusted him and he was immensely disliked. But because his last name was on the outside of the building in three-foot-high block letters, people put up with him by doing just the minimum amount of work to stay gainfully employed because they needed a paycheck.

Then one day it all exploded in his face. King Ironfist came into the warehouse. He saw a college student lounging against a pallet of boxes drinking a can of soda and reading the newspaper. He raced up to him and yelled, "Hey, kid, how much money do you make a week?"

"One hundred seventy-five bucks," the kid replied.

King Ironfist reached into his wallet, pulled out $175, and said angrily: "Here's a hundred and seventy-five dollars, hit the road. I don't ever want to see you back here again!"

We all watched this in horror. Next, King Ironfist grabbed the shipping and receiving supervisor and said, none too politely, "I ought to fire you too; when in the hell did you hire that lazy bum?"

"I didn't," retorted the supervisor. "He's a driver for Glenwood Freight and he was just waiting for us to finish loading his truck!"

After King Ironfist stormed off, all of us in the warehouse crew laughed until our sides ached; the joy we felt was substantial.

Reflecting back on that incident many years later, I see it in a completely different light—a darker and more disturbing light. King Ironfist had created such a demoralized and demotivated culture throughout all departments that when news of the incident spread, everyone was exhilarated over his screw-up. Employees taking delight in their president's failures is the epitome of shattered trust and a complete absence of collaborative relationships.

How much better the outcome could have been if the president had built a culture of mutual trust. Then, if he perceived a problem, he could have held a 30-second private discussion with the shipping and receiving supervisor—the one closest to the action—and quickly learned who the "slacker" really was.

Trust is the road over which all of your other collaborative leadership behaviors will ride; it is the key to your integrity and ability to be a respected leader instead of an immensely disliked commander-and-controller getting minimum effort from all your employees.

Trust and collaboration are so tightly interwoven in the workplace it is hard to separate them. Lack of mutual trust blocks genuine collaboration just as lack of collaboration thwarts building trust and support. Without mutual trust, collaboration is unworkable as people shut down, pull back, and hesitate to engage; likewise, without opportunities for meaningful collaboration, employees do not have the chance to work together, to engage, to build and honor commitments, and thus earn each other's mutual trust. Done well, the two actions will reinforce each other in building superb working relationships; done poorly, the two will reinforce each other in tearing down relationships.

DEFINING MUTUAL TRUST

Jack Welch, former CEO of GE, reinforces our sidebar definition of trust with some insightful comments:

> Trust is the confident belief in, and reliance on, the integrity of another person's words and actions.

> Trust fritters and dies two ways. First, when people aren't candid with one another; when they sugarcoat tough messages; when they use jargon and baloney to purposely make matters obscure and themselves less accountable. The only way to get candor into an organization is for the bosses to identify it as a top value, and consistently demonstrate it themselves. The second trust killer is when people say one thing and do another. Trust, ultimately, isn't very complicated. It's earned through words and actions instilled with integrity. Say what you mean and do what you say![1]

Mutual trust is reciprocal (you've got to give it to get it), and it is created incrementally (step by step over time). Trust is earned slowly, but it can be lost quickly. Mutual trust and support is the basis for effective leadership. When leaders attain high trust with team members, everyone is at the peak of their power and in a perpetual state of readiness for unencumbered interactions—the kind of interactions that produce extraordinary organizational achievements.

THE 3-Cs OF TRUSTWORTHINESS, YOUR PERSONAL KEYS TO TRUST BUILDING

As noted previously, mutual trust is the bedrock for the road over which an organization's success rides. The bedrock includes three blended elements: competence, character, and communication. We will refer to these as the "3-Cs" of trustworthiness, which will be shown to be essential to your ability to build mutual trust with and support of others.

Competence: Belief among People That Each Can Do What They Claim Is Their Ability to Do

1. Trust *builders* include: seeking and acting on inputs, encouraging responsible risk taking, extending growth opportunities, increasing responsibility and authority, encouraging people to monitor their own progress toward goals. By doing any of these, trust is exhibited in another's competencies. As a consequence, self-confidence is nurtured and trust is given back in return.
2. Trust *breakers* include: micromanaging, failing to delegate, keeping people pigeonholed in routine assignments. As a result, people aren't given opportunities to shine and demonstrate their latent capabilities. They feel ignored and deprived of opportunities to grow and develop. Trust erodes.

Character: Belief among People That Each Will Do What They Say They Will Do

1. Trust *builders* include: keeping promises/commitments or renegotiating them as needed, keeping confidences, acting

consistently regardless of the person or situation, collaborating freely, crediting others. These actions set one's reliability and credibility at a high level and enable trust to thrive.

2. Trust *breakers* include: avoiding responsibility, passing the buck, engaging in self-serving actions at the expense of others, making excuses or blaming others when things don't work out, being unethical, misrepresenting data/information to make one look good. These off-putting character behaviors ruin trust, collaboration, and people's willingness to work at their best because personal integrity is absent.

Communication: Belief among People That Each Will Share Information Openly and Honestly

1. Trust *builders* include: listening to and valuing what another says even when not in agreement, admitting mistakes, giving and seeking constructive feedback, attacking issues without getting personal, explaining reasons behind requests and decisions, providing ongoing status reports. Trust grows when communication is based on sincere give-and-take without fear of personal reprisals.
2. Trust *breakers* include: betraying sensitive information, gossiping, sending mixed messages to obscure true feelings, using doublespeak, holding hidden agendas in discussions, jumping to conclusions without checking the facts. When people experience these communication breaches, relationships are damaged and an unhealthy culture of mistrust is cultivated.

Your trustworthiness grows or degrades based on the degree to which people perceive *trust building behaviors* as your norm within the 3-Cs versus *trust breaking behaviors*. Note the word "perceive." Trustworthiness, then, is subject to people's "perception" of the 3-Cs—competence, character, and communication—in one another. Perception based on many personal interactions will be more accurate than perceptions based on only a few—or even no—contacts. Still, what each person "perceives as real" is that individual's reality. To stop false perceptions from taking root, everyone in your organization has to take responsibility for setting and maintaining high personal standards of trust building across their competence, character, and communication.

A MODEL FOR BUILDING AND SUSTAINING MUTUAL TRUST

As emphasized before, trust is a mutual concern. Each person in the relationship has to be interested in what happens to the other. The model, shown in Figure 4-1, is one I've created to map out a process that visually reveals how you can build and preserve mutual trust for the good of all parties involved. It also provides you valuable insight into how to avoid the pitfall of mistrust. The principles and steps of our model are easily understood and translated into action. When employed by you and many others at different levels in your organization, the results—in terms of developing leadership credibility, employee engagement, team collaboration, and organizational performance—will be noticeable. Remember, this is the core of your collaborative leadership activities.

There are the two primary components to our model for building mutual trust. They both interact in a constantly reinforcing cycle to build and preserve the veracity of a trustworthy and trusting relationship among you and others.

Component number one is made up of the essential 3-Cs (competence, character, and communication), which gives you the bedrock—the reputation for being trustworthy. Because of your standing as a trustworthy individual, others believe: you *can do* what you claim is your ability to do, you *will do* what you say you will do, and you will share information *openly and honestly.*

Being known for the 3-Cs gives you the eminence to build and nurture trust with others on two levels. One is where the current level might be strained or even damaged. If that's the case, engage the relationship at point A. The 3-Cs also will allow you to be accepted as a respected leader or teammate when working together with other trustworthy individuals where the past level of trust has been good and you want to sustain or expand that relationship. When that's the situation, engage the relationship at point B.

Figure 4-1. A Model for Building and Sustaining Mutual Trust: A Reinforcing Cycle

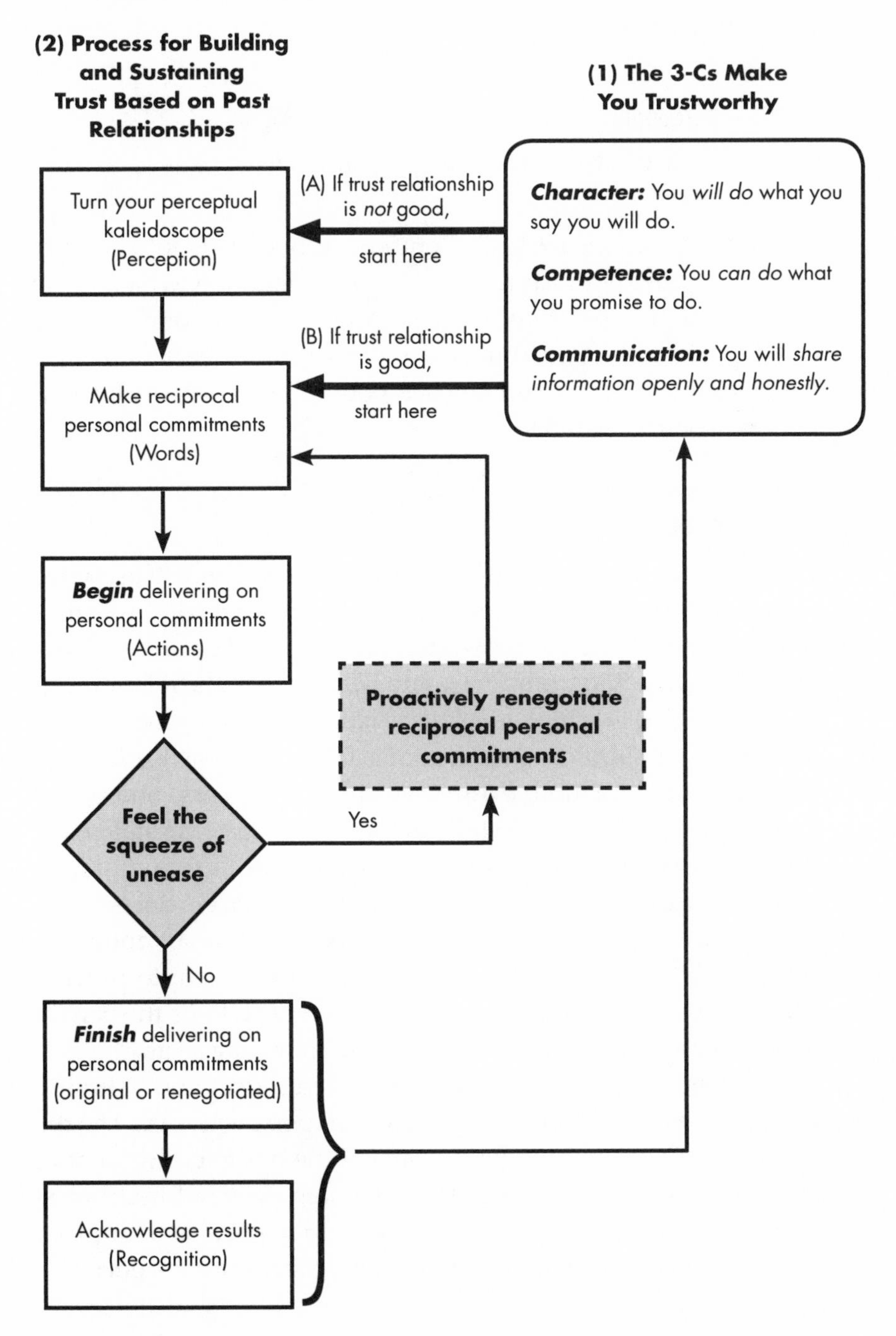

The second component is the actual behavioral process for creating, sustaining, and expanding mutual trust as well as adding to your 3-C reputation. It is comprised of a set of five actions that are your trust building enablers.

Perceptions → Words → Actions → Recognition,
and (when necessary) Renegotiation

The full model is cyclical. Component one feeds into component two, which in turn reinforces component one, which again bolsters component two—and so around and around we go earning and building mutual trust. Let's examine each of the five behavioral actions so you can see how the full process for perpetuating trust operates.

Turn Your Perceptual Kaleidoscope (Perception)

This has to be the starting point for building genuine trust when the trust level in the relationship is tense or severely broken. If someone in a relationship doesn't take the risk of trusting, of seeing the relationship from a new and better angle, the relationship remains stalled at a high level of caution and suspicion.

In superior-subordinate situations, it has to be the leader who takes this initial action. With colleagues and peers, one of the parties has to be willing to take the risk and break the stalemate. A good way to turn your kaleidoscope and take the initial step in building mutual trust is to *assume positive intent.* At times when trust is lacking, one person assumes the worst about another's actions and motives. Instead, assume positive intent to prevent a knee-jerk opinion based on rumors or hearsay. Give the benefit of the doubt. Things aren't always what they may seem.

For some people with deeply ingrained perceptions, turning one's *perceptual kaleidoscope* to get a fresh perspective may be the hardest step to take. Yet this is a must if there is to be any chance of improving a relationship. This first step does not build trust by itself. It's only an entry point to begin the process of improving rapport and affiliation. The remaining steps are the ones that will strengthen mutual trust building in relationships and burnish people's 3-C reputations.

Make Reciprocal Personal Commitments (Words)

Resetting your perceptions so they are more favorable regarding another person's—or group's—desires or intentions opens the door for engagement. If past relationships have proven to be trustworthy, then no kaleidoscope turning is required; step two will be your starting point.

Irrespective of how you got to step two, it quickly puts trust building into a constructive stage—the stage of verbal interaction and personal exchange, where *reciprocal personal commitments can be made*. So at step two you will be on the high road, either beginning to build mutual trust and support out of a previous negative situation or sustaining and enriching one based on a previous positive encounter. Regardless, the actions are the same.

It is here that questions can be raised, mutual expectations processed, boundaries agreed to, deliverables committed to, and timelines set. By doing this, opportunities to earn and uphold each other's trust and respect are created because collaborative dependencies have been formed.

With reciprocal commitments, you're now counting on me to help you be successful, and I'm counting on you to help me be successful in achieving our shared goal. Furthermore, the team is counting on both of us to be successful. If we deliver on our mutual promises, my success is your success, your success is my success, and our mutual success is the team's overall success.

This is no time for dishonesty! All 3-Cs, among all parties involved, are being tested in this second step. Anyone acting sporadically, inconsistently, or insincerely with respect to any of the 3-Cs is undercutting the integrity of the trust building process. Knowingly overcommitting, in order to impress others with how valuable you are and what great things you can do, when you know you can't deliver, will break *competency trustworthiness*. Being unethical, skirting the issue instead of directly addressing it, distorting data/information to make you look good, will destroy *character trustworthiness*. Sending mixed messages, expounding half-truths, or holding hidden agendas will shatter *communication trustworthiness.*

However, with authentic, genuine interactions taking place, we are making reciprocal personal commitments; we are learning about each other; we are collaborating; we are putting our trustworthiness on the line; we are setting the stage for mutual trust; we are creating hope. We are doing a lot of talking, negotiating, and committing; but we are not there yet.

Deliver on Personal Commitments (Actions)

Now it is time to deliver; now it is time to "walk the talk." During step three the parties carry out what they have promised to do for each other. Success in this step puts a solid gold brick in the road of trust. Each and every time commitments *are made and kept* among team members, another solid gold brick is laid down. For some relationships this may be a routine activity; but for others this may be the first time a gold brick was laid down in that relationship.

The Squeeze of Unease (A Warning Signal)

Even with the most careful, honest, and sincere setting of mutual commitments in the *words* phase, there is potential for many things to go awry during the *action* phase of delivering. Not living in a perfect world, circumstances change. Sometimes the change is significant enough that it rocks the stability of the initial commitments made. Unforeseen problems, additional complexities, decisions beyond one's control, changing priorities, all can play a hand in derailing our ability to meet some portion—or even all—of the personal commitments established with others in the second step.

While there is no "silver bullet" assuring success in situations like this, the action that comes closest is straightforward: *heed the squeeze of unease.* The "squeeze of unease" is the realization, sometimes physically felt—like anxiousness, fear, a twinge, a headache—that a disruption of personal commitments is likely to be a reality. It arises when you hear, see, or read new information that puts the original commitments at risk; or when you find it difficult to live up to personal commitments because you are forced to cope with unexpected extenuating circumstances. Here are two examples of the "squeeze of unease" in the middle of delivering on personal commitments.

1. The HR manager has made a personal commitment to you, a division VP, to fill several key vacancies in your organization within six weeks. As that deadline draws near, the HR manager realizes that the selection and hiring process is not yielding results as quickly as anticipated and she is in serious jeopardy of not being able to deliver on her commitment.
2. You have made a personal commitment to another department head to help orient and mentor his newly hired manager for the next three months. Recently, on top of everything else, your own leader has given you a major crisis task to manage. This new assignment is very time consuming and requires your full focus. Your mentorship has taken a back seat and left your mentee feeling confused and abandoned, as you have had no interactions with him for a month now.

Whenever any person in a relationship of mutual commitment feels the "squeeze of unease" for whatever reason, three options are available: (1) do nothing, (2) assure the other person(s) that all is well and try to fix things behind the scenes, or (3) reopen communications and proactively renegotiate a new set of commitments taking into account the changed circumstances.

Option 1: Do nothing. This is foolish and risky on two counts. If things don't work out and you don't deliver, your competency trustworthiness takes a hit. And keeping the people in the dark who are depending on your deliverables creates anxiety and uncertainty for them. We all tend to have too much to do at work, and in trying to do too much we often short-circuit our communications and renegotiations. While the trust erosion may be an unintended consequence of trying to do too much, it is just as real. Actually, by trying to show how full of "can do" and "will do" you are, you take a major risk of demonstrating neither.

Option 2: Saying all is well when it isn't. This is communication with a hidden agenda; it is dicey; it is dishonest. If you cannot fix things, the result is the same as above, but magnified by the dishonesty involved. Your character, communication, and competency trust all are spoiled.

Option 3: Proactively renegotiate. This is your only true option because it builds trust. Even if you ultimately end up delivering less than what was originally committed to, trust is built since mutual expectations were reset.

Proactive Renegotiation (It Must Be Done)

As soon as you receive signals that produce a "squeeze of unease" regarding delivery of your personal commitments, you must communicate with the appropriate people and identify your "squeeze" so mutual action can be taken before disruption occurs. It may be as simple as a phone call or a couple of e-mails, or it may require a series of difficult face-to-face sessions to renegotiate new expectations. But do whatever it takes when trust is at risk.

Encourage, even demand, that others come to you when they feel a "squeeze of unease" regarding their ability to do what they promised to do. Earn, and hold dear, your reputation as a giver of constructive feedback, a problem solver when people do come to you with issues. Mark Murphy conducted one of the largest studies ever on the topic: what makes employees trust their boss. He discovered that the *extent to which leaders respond constructively* to employees who bring them work-related problems is the biggest driver of employee trust.[2]

Proactive renegotiation is smart business. It affords you and the others more choices and more control over the alternatives. You have a greater chance of working out a satisfactory solution under these conditions, as opposed to trying to do so in a cauldron of *forced* renegotiations after things have already exploded.

Therefore, as soon as a person (or team of people) feels the grip of the "squeeze of unease" vis-à-vis personal commitments, it must be dealt with immediately. It cannot be put off as someone else's responsibility. Setting an explicit ground rule that puts a premium value on proactive renegotiation as a fundamental part of any collaborative relationship is a good idea. This wards off doubt and anxiety and is the key to building trust!

Acknowledge Results (Recognition)

Every time you and another person or group is successful in delivering on your mutual commitments—even if the final outputs were less than what was originally committed to but were proactively renegotiated—you should do something to signify your achievement. That means acknowledging the results.

Making thank-you phone calls, sending e-mails for a job well done, going to lunch together, meeting after work for a drink and informal socialization, are all simple ways to reinforce accomplishment and show appreciation for each other. The strength of this last step is its symbolic power. What better way to grow and maintain trust within or between work teams than by acknowledging mutual success?

ELIMINATE THE SUPER TRUST DESTROYERS

Earlier, a number of trust breakers were noted under each of the 3-Cs. However, three actions stand out as the super destroyers. Any one of these three will ruin your character, competency, and communication trust with equal vengeance and render you untrustworthy and incapable of collaborative leadership or being a collaborative worker.

Breaking Original Commitments without Feedback

Nothing will ruin your credibility faster than breaking your original commitments without any feedback or earnest attempt to renegotiate new ones. Breaking even seemingly insignificant commitments can blow your trust relationship sky high if the other person was seriously counting on you to deliver. Through your action, or lack of such in this case, the message you are sending is: "I don't walk my talk." Your action message overrides everything you say in words; very soon, people stop listening to what you say.

Shooting the Messenger Who Comes with a Squeeze of Unease

A team member coming to you at the earliest possible time, communicating a "squeeze of unease" that will alter his or her ability to meet original commitments, is providing a great service to you and the team. However, when team members fear your reprisals to "bad news," they stop delivering it. They dilute the truth; or worse yet, only tell you what you want to hear, leaving out all negative information. Trust development is impossible because no team member will ever be in a state of readiness for unencumbered interactions with you, or anyone else, who "shoots messengers bearing bad news."

CYA: Covering Your Ass

Team members who avoid taking responsibility for their commitments when things go awry—by passing the buck, shading the truth, tap dancing around the facts, finger pointing, blaming, attacking others, making excuses—are killing their trustworthiness. CYA behavior is egotistical, self-serving preservation at the expense of other team members. It blackens all of your 3-Cs, paints you as an egotistical conniver, and wrecks the spirit of team collaboration.

NEVER COMPROMISE YOUR TRUSTWORTHINESS; NEVER STOP BUILDING TRUST

It is hard to say "never" with respect to any leadership behavior, but if there ever was one behavioral pronouncement where "never" applies, it is the one noted in the section title. You just can't compromise your trustworthiness, take it for granted, or become lax in working on building trusting relationships, ever! If you do, then the centerpiece of your collaborative leadership behaviors will crumble and all the others will cave in on top of it. Let's look at a few comments by others to underscore my point.

More than four decades ago Douglas McGregor, in his influential book *The Professional Manager*, described the importance of trust, how fleeting it can be if not diligently fostered and maintained, and how it intertwines with open communications.

> Trust means "I know that you will not—deliberately or accidentally, consciously or unconsciously—take unfair advantage of me." It means "I can put my situation at the moment, my status and self-esteem in the group, our relationship, my job, my career, even my life in your hands with complete confidence"...
>
> Trust is a delicate property in human relationships. It is influenced by your actions more than your words. It takes a long time to build, but it can be destroyed very quickly. Even a single action—perhaps misunderstood—can have powerful negative effects. It is the *perception* of the other person and of his actions, not the objective reality, on which trust is based....
>
> Mutual trust and open communications are closely interrelated. Open communications help prevent misperceptions of actions, but inconsistencies between words and actions decrease trust....The effective performance of a managerial team is in ...a sense a function of open communications and mutual trust between all members including the leader.
>
> My view then, clearly influenced by my values, is that one of the fundamental characteristics of an appropriate managerial strategy is that of creating conditions...for a climate of mutual trust, mutual support, respect for the individual, and for individual differences. Only in such a climate can the latent tendencies toward self-actualization find expression.[3]

In their book, *Leaders: The Strategies for Taking Charge,* Warren Bennis and Burt Nanus cite trust as a key element of effective leadership: "Trust is the emotional glue that binds followers and leaders together. The accumulation of trust is a measure of the legitimacy of leadership. It cannot be mandated or purchased; it must be earned. Trust is the basic ingredient of all organizations, the lubrication that maintains the organization."[4]

Simply put, trust means confidence—confidence that others' actions are consistent with their words, that the people with whom you work are concerned about your welfare and interests apart from what you can do for them, that the skills you have developed are respected and valued by your teammates, other leaders, and the larger organization, and that who you are and what you believe truly matter in the workplace. Trust becomes pervasive when—and only when—the organization prizes trust as a key operating principle and it is role modeled and supported by senior leaders. By modeling the values of trustworthiness and trust building, senior leaders provide a benchmark for all employees. In addition, senior management communicates expectations clearly so that fear, defensiveness, and reactive behaviors are replaced with proactive, responsive actions that improve quality, reduce costs, and better meet customer needs.

James Kouzes and Barry Posner are two authors who have written widely on the topic of leadership. In their book *A Leader's Legacy,* they provide commentary on the subject of trust.

> So, if you want the best relationships and outcomes, you have to trust. And you have to understand that in the game of trust, it's the leaders who have to ante up first. This means taking a lot of time to build relationships. It means listening carefully to others. It means getting to know about their capabilities, needs, and aspirations. It means talking about values and being clear about norms—for example, what's acceptable and/or unacceptable in terms of how people treat each other, regardless of their place in the organization. It means being on the same page about performance standards, customer expectations, and about why what we do matters.[5]

OUR MODEL FOR BUILDING AND SUSTAINING MUTUAL TRUST

From Douglas McGregor in 1967 to contemporary writers of today, the topic of trust remains at the forefront of leadership behaviors. Its importance and power in the development of collaborative leadership skills and in building highly respected businesses and institutions cannot be overstated.

Trust cannot be built unless there are interactions between two or more people. Forming relationships is a must to building trust. If relationships have been good in the past, this history greases the skids for another successful encounter based on mutual trust and support. And if all goes well—as it most likely would—everyone's status as trustworthy, trusting individuals grows.

However, when trust is damaged, the challenge is more difficult. Someone must turn their perceptual kaleidoscope and look at the other person or team from a new, more positive viewpoint. Initiating this action will fall to you, the leader. And when you do it, you are taking a risk. You are making yourself vulnerable. You expose yourself to the consequences of the actions of others upon whom you become dependent. No longer are you fully in control. You will have doubts as you collaborate to renew a faltering relationship. A hundred questions like these will haunt you until the task is finished: Is the other party really committed to the task as much she claims? Will the deliverables be met as agreed to? Will I be communicated with as soon as the other party feels the squeeze of unease? What happens to me if things really get screwed up?

But if you don't trust others, what happens then? You will never build trust. You will, whether you like it or not, have to be a micromanager—telling, selling, checking, and pushing. You will be overburdened carrying much more of the workload because you don't trust others to do their part. You will get compliance from your people just as King Ironfist did, with your people doing just enough to stay employed. And the more you don't trust others—the less faith and confidence you express and demonstrate in them—the less they come to trust you in return. You will hold back the growth and development of others. And finally, the potential health problems from the stress, pressure, and burnout you are sure to feel working in the environment you created should not be discounted.

Everything written about in this chapter ties back to and revolves around Figure 4-1; it is your everyday, "how to" model for creating, building, and expanding trust between you and others. It all starts with the 3-Cs: competence, character, communication. With those in place, so you are recognized as a person who consistently demonstrates trust building behaviors, there are two places to engage others in the left side of the model. Remember, however, once you enter the second phase ("making reciprocal commitments"), the process from then on is exactly the same.

The more times you turn your perceptual kaleidoscope and are willing to attempt to rebuild mutual trust and support by engaging others in working through the model, *the more trusting you become*! The more times you complete the full trust cycle with different people, the larger your personal trust bank account grows. That is, more and more people will have the confident belief in and reliance on *your* words and actions, and *the more trustworthy you become*!

I'll conclude by sharing three of my favorite reminders about the operation of trust in building strong and lasting relationships:

- When you make a commitment, you only create hope; when you keep a commitment—even if renegotiated—you build trust.
- Trust formula: $T = CM/t$
  Trust = commitments met over time.
- Trust is a must or your leadership is a bust.

BUILDING MUTUAL TRUST: KEY LEARNING POINTS AND WHAT I WANT TO DO DIFFERENTLY TO IMPROVE

My Key Learning Points from Chapter Four:

What I Want to Do Differently:

READER REFLECTIONS AND APPLICATION ACTIVITIES FOR CHAPTER FOUR

Reflections

Reflect on the following points for your own "action research" around trust. Think of two important people in your life who are part of your relationships network: one with whom you have a positive relationship and one with whom the relationship is not so positive.

1. What differences did you discover regarding your assessments of competence, character, and communication for each relationship?
2. What was different for you emotionally (i.e. how did you react/respond) in each relationship?
3. What was different about the quality of the interactions?
4. What lessons can you take away from each relationship assessment (to start, stop, or continue doing some behavior or activity) in order to strengthen your own trust building behaviors?

Leader Application Activity: Building Mutual Trust with an Individual (30 to 45 minutes)

1. Have the teammate read Chapter Four and come prepared with answers to two questions in advance of meeting with you:
   - What did you feel were the most important learning points in the chapter?
   - Why were these key learning points for you?
2. Open the session by letting the teammate share his or her views on the two prework questions above. Actively listen, understand.
3. Share what you found important in the chapter and why these points were key to you.
4. Use the "Building Mutual Trust" process model as a centerpiece and talk through each step, sharing viewpoints and perspectives.
5. Test the Figure 4-1 model in real time by making several mutual, reciprocal commitments regarding a particular task.
6. Follow the model using trust building behaviors—as you attempt to fulfill your mutual commitments and renegotiate commitments as necessary.

Leader Application Activity: Building Mutual Trust with Your Team (45 minutes)

1. Have all teammates read Chapter Four and come prepared with answers to two questions in advance of the meeting:
   - What did you feel were the most important learning points in the chapter?
   - Why were these key learning points for you?
2. Open the session by asking teammates to share their views on the two prework questions. Actively listen, understand. Note responses on a flip chart.
3. Share what you found important in the chapter and why these points were key to you. Add these to the flip chart.
4. Probe the team to discover "what we do well" and "what we need to do better" regarding trust behaviors within our team (and/or with HQ, operations, staff groups). Be specific, use examples to clarify pluses and minuses.
5. Use the Figure 4-1 process model as a centerpiece and talk through each step, sharing viewpoints and perspectives.
6. Get consensus that the "Building Mutual Trust" model will become a shared operating practice across the team. Initiate the model's use by applying it to a current situation within the team.

NOTES

1. *Business Week,* August 14, 2006, 88.
2. M. Murphy, *Hundred Percenters: Challenge Your Employees to Give It Their All and They'll Give You Even More* (New York: McGraw-Hill Companies, 2010), 198.
3. D. McGregor, *The Professional Manager* (New York: McGraw-Hill Inc., 1967), 78, 163–164.
4. W. Bennis and B. Nanus, *Leaders: The Strategies for Taking Charge* (New York: Harper-Row Publishers, 1985), 153.
5. J. Kouzes and B. Posner, *A Leader's legacy* (San Francisco: Jossey-Bass, 2006), 76.

CHAPTER FIVE

DECISION MAKING

The Range of Options and the Forces Impacting Your Decision Choice

CHAPTER OBJECTIVES

- To present a framework for classifying and understanding the range of decision-making options available to the work group manager or task force chairperson
- To highlight the characteristics of each of the eight options
- To introduce and describe five noteworthy forces impacting the selection of a particular option

INTRODUCTION

This little rhyme, often recited by one of my college professors to keep us grounded in the realities of decision making once we entered the "real world" of business, has stayed with me to this day. It does sum up the common dilemmas of decision making.

> You can and you can't,
> You shall and you shan't;
> You will and you won't;
> You'll be damned if you do,
> And you'll be damned if you don't.

While you may not always be "damned if you do" and "damned if you don't," for complex decisions it often seems that way. Moreover, when facing what seems like a tough or unpopular no-win situation, people often retreat from making *any* decision.

While Chapter Five will pinpoint eight decision options available to you in a decision-making situation, it is important to note that a frequently used decision method is the one just mentioned—"decision by not deciding" or "decision by default." R. K. Mosvick and R. B. Nelson elaborate on this point:

> Dr. Marcus Alexis, former chairman of the Interstate Commerce Commission, estimates that as many as 50 percent of the important organizational decisions end up being decision by default. Experienced managers are all too familiar with this phenomenon. They have observed too many potentially good decisions become strangled by indecision, consigned to some bureaucratic limbo, or overtaken in one brilliant move by a competitor. Decision by default continues to thrive as one of a number of irrational decision methods too vast to catalogue here. It is best seen as a sharp reminder to all that a decision by a nondecision is still a decision.[1]

The eight decision-making options to be covered in this chapter are all proactive. The decision avoidance approach was mentioned here simply to acknowledge its existence. It is obviously not a desirable option since in fact it is decision abdication. Indeed, decision by nondecision means you are surrendering decision-making influence to "the four winds," "chance," "fate," "the system," or "the powers that be."

UNDERSTANDING THE EIGHT BASIC DECISION OPTIONS

Any leader has a deck of eight decision options to draw from when entering into a decision-making situation. They are all viable, legitimate, and useful ways to make a decision. The danger in reading a chapter such as this is making value judgments about various options in the array and deeming some to be better or more mannerly than others. However, doing so is a mistake. As you read and learn about each of these eight decision approaches, it is important that you keep the following cautions in mind at all times.

> *Do not judge one method as being better than another.* Each alternative presented has its time and place. Each can produce positive as well as negative consequences, depending on how it is applied within the context of a particular situation.

The tree diagram shown in Figure 5-1, then, presents a simple framework for classifying the eight rational decision options available to any leader. Regarding the tree diagram itself, you read it by moving from left to right. Along the top are three boxes defining each of the three classification breakdowns that guide you to a particular decision option. Footnotes along the bottom clarify key points regarding different aspects of the decision classification process.

As you move from left to right on the diagram, the first division is made according to the two basic decision-making **approaches**—*autocratic and shared*. Next, these broad approaches are refined to reveal the two decision **types** within each one. The autocratic approach contains the *pure autocratic* (PA) and the *consultative autocratic* (CA) types. For the shared approach, there is the *partial group* (PG) type and the *whole group* (WG) type. Finally, our framework delineates a total of eight specific decision-making **options** (two options per type). Each of the eight options will be described more fully in the next two sections to give you an understanding of their mechanics, strengths, and drawbacks.

REVIEWING THE FOUR AUTOCRATIC OPTIONS

The four autocratic options are available to you whenever you hold final decision-making authority, through your position power, regarding the issue being processed. However, if you are a chairperson, committee leader, or task force manager, or are in

Figure 5-1. A Method for Classifying Decision-vMaking Options

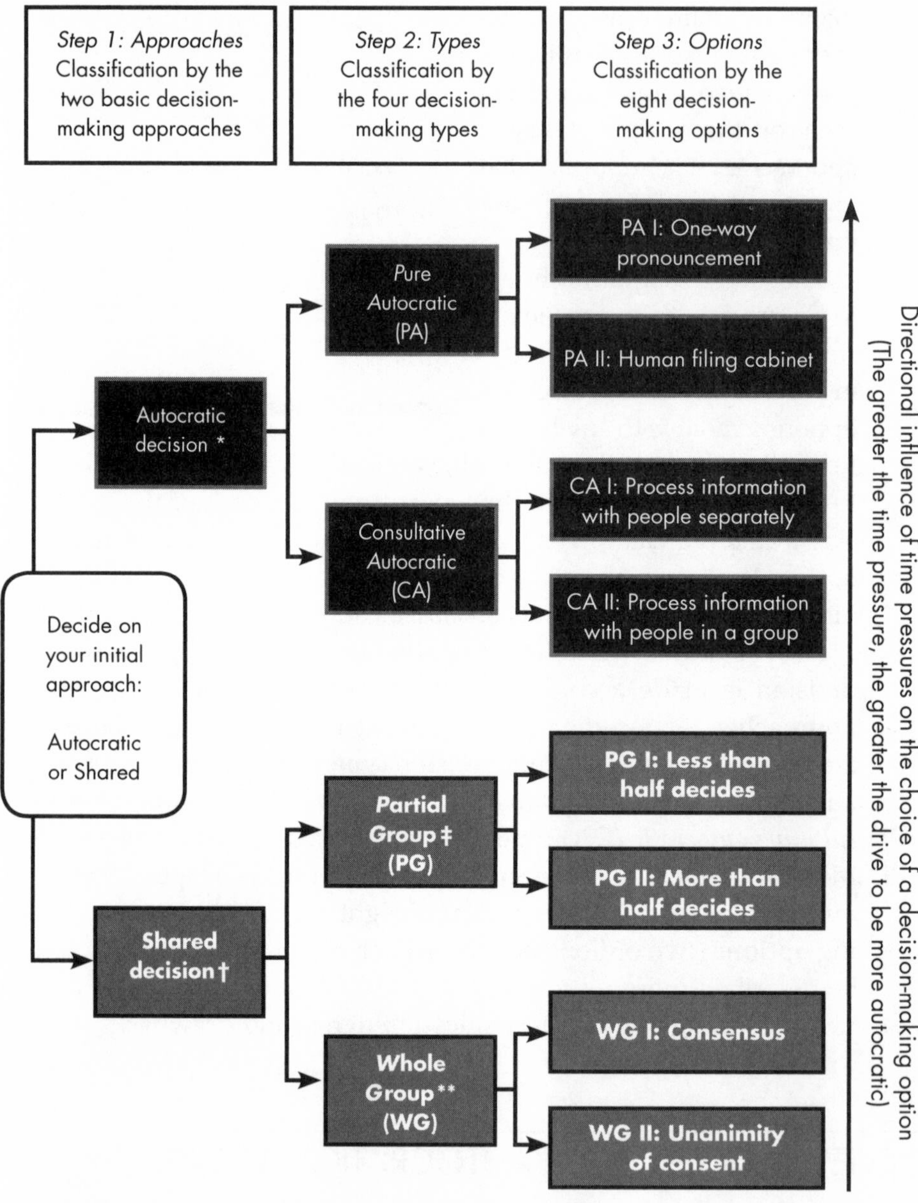

* The *autocratic* decisions approach contains four options in which the *authority* to make the final decision *rests with one individual.*

† The *shared decisions* approach contains four options in which the *authority* to make the final decision *rests with more than one individual.*

‡ The two Partial Group options (*PG I and PG II*) are options whereby part of the group makes a decision for the whole group. *PG options always carry the burden of being win/lose* unless carefully facilitated.

** The Whole Group options (*WG I and WG II*) are options whereby the entire group reaches a joint decision fully supported by all members. *WG options are win/win.*

some other leadership position—*but without final decision-making authority*—you will not be able to use these four autocratic methods. Instead, you will have to facilitate decisions using one of the four alternatives from within the *shared* categories.

Pure Autocratic I: One-Way Pronouncement

A baseball umpire working the game behind home plate is the perfect example of the pure autocrat. As Bill Klem, a National League Hall of Fame umpire from the 1920s through the mid 1940s, put it so eloquently regarding balls and strikes: "They ain't nothin' until I call 'em."

With the pure autocratic (PA I) option, you or some other person with final decision-making authority simply makes a decision and announces it to others. The "others" have had no part in thinking through the final decision announced. PA I decisions sound like these: "I've decided I'll write the memo to Ursula explaining our $3,500 cost overrun." Or: "I want you to have a full analysis of the problem on my desk by Tuesday morning so I can defend our interests in the policy meeting that afternoon."

A group can have a PA I decision transpire and not even know it because it is so subtle. This occurs when someone, without any discussion, acts immediately on his or her own proposal and the group goes along with it (this is called a self-authorized decision). For example, a teammate grabbing a marker and going to a flip chart saying, "I'd better start writing this stuff down," and then starts noting the comments. Or: "My uncle is the manager at the Silver Lake Country Club; it's perfect for our party, so I just went ahead and booked it for us."

This pure autocratic option and its variant shown next are both useful when centralized decision making and coordination are required during an emergency or crisis.

Pure Autocratic II: The Human Filing Cabinet

You use this option to obtain information from other individuals without sharing the problem or need for the information—hence, using the people as human filing cabinets.

For example, regarding a budget cut for your department, you would speak to an employee privately and might say, "Marty, I would like you to get me a copy of your unit's year-to-date spending versus target for the last six months, itemized by budget account numbers." After doing this with all your direct reports, you would process the data alone, determine which accounts are to be cut by what amount from each unit's budget, and announce the revised budgets to your staff as a "done deal."

Consultative Autocratic I: Processing Information with People Separately

You use the CA I option to share the problem, *one-on-one*, including some of the relevant background issues. You may or may not include your current inclinations. Then you would seek each individual's ideas, proposals, advice, or counsel as inputs for you to utilize as you decide. A variation of the CA I decision type is to talk to people one-on-one not to gain insight, but rather to convince them to back a decision you intend to make or to line up support to stop a decision some other person or group wants to make.

For example, regarding your clothing store's image, you could say: "Chris, I'm asking all our store employees for ideas to help me fix our store's poor image problem once and for all. What ideas do you have on how we can impress customers when they enter our store?" After talking to each store employee one-on-one and gathering individual responses, you would process the information alone, decide what actions would be taken to improve the store's image, and announce the plan to everyone.

Consultative Autocratic II: Processing Information with People in a Group

You make use of this method by sharing and processing the problem with others *in a group, but you retain final decision-making power for yourself.*

To use the CA II option with integrity means you enter into the group debate and discussion with an open mind and make clear that while you retain the final decision-making rights, you are willing to be influenced and you have no predetermined outcome in mind. Let them know everything discussed will be considered by you as you make the final decision.

The big advantage of CA II over CA I is the collaborative interplay among the members in the group. The synergy, enthusiasm, and power of multiple ways of thinking cannot be overlooked. When announcing the final decision, if you can share some examples where the group's inputs influenced your thinking in arriving at the final decision, so much the better. This will help generate commitment to what you have decided.

For example, bringing your direct reports together, you might say: "I would like your thoughts and feelings about the proposal to combine the management development and technical development operations into one department. After we mutually process your views, pro and con, I will think about everything, make my final decision, and present it to the organizational restructuring committee tomorrow." Once you collected your teammates' inputs from the roundtable, you would seriously reflect on them along with other pertinent information and considerations, make your decision, and review it with the restructuring committee the next day. After that, you'd inform your direct reports of the decision you presented to the committee, your rationale for it, and specifically comment on how your team influenced your thinking as you arrived at your final decision.

Stanley Kramer's experience in making his 1963 classic movie *It's a Mad, Mad, Mad, Mad World* illustrates the consultative autocrat in action. The movie was a comedy on a grand scale and featured a multitude of world-famous comedians, including Buster Keaton, Joe E. Brown, Milton Berle, Sid Caesar, Jerry Lewis, Buddy Hackett, and Jonathan Winters. When asked how he maintained control over these rather egotistical performers, each wanting to do his own shtick, Kramer admitted the first week of rehearsals was a staring contest between the comedians and the serious-minded producer/director (himself). Then Kramer held a "deshticking session." He gave the comedians the freedom to make contributions about routines, setups, and even script changes, and he told them he would listen and discuss proposals with the people involved. But he would have the final say as to what changes were actually made. Kramer noted that once everyone understood and accepted the process, "we were in business."

REVIEWING THE FOUR SHARED OPTIONS

The *shared decisions* approach contains four options in which the *authority* to make the final decision *rests with more than one individual.* With all four options, you'll be highly involved in all the information processing, but you will not be the final decision maker.

Partial Group I: Less than Half Decide

Using this alternative allows you to share the problem with the group and facilitate a discussion to process information about the issue. At some point a decision is made by less than half of the group for the whole group.

PG I can be a very constructive decision-making approach when it becomes apparent, either before or during the discussion, that a group of individuals—comprising less than half the total group—possesses the knowledge, skills, and expertise to make a quality decision that the entire group can support. Realizing this, the majority defers final decision-making authority to the minority group.

However, PG I can be filled with pitfalls that turn it into a nonconstructive, politically explosive win-lose process. One example of using the process in this way is "railroading," wherein a coalition composed of less than half of the people present emerges. Then, because of their rank, influence, expert knowledge, charisma, political power, verbal skills, or some other factor, they are able to drive the final decision in their favor even though the members of the majority do not support it.

Partial Group II: More than Half Decide

Using the majority rule option allows you to share the problem with the group and facilitate a period of open discussion. At the conclusion of the discussion period *you poll everyone, and if a majority shares the same opinion, the decision passes.*

More formally, at the end of the discussion period you would state clear-cut alternatives and solicit votes in favor, votes against, and abstentions, with the outcome determined by majority rule. However, while voting may seem straightforward, you should be aware of some potential snares before facilitating this process.

First, you and the group need to agree on the definition of *majority* before the vote is taken. Do you mean a simple majority, with 51 percent concurrence required to carry the decision? Or do you mean a two-thirds majority, with 67 percent needed before the decision passes?

Second, voting plays up differences among members and makes taking sides unavoidable. Voting tends to split the group into opposing factions, which encourages members to pay more attention to the arguments of people on their side than to actively listen to the arguments of the opposition.

Third, decisions made by the voting process may turn out to be poorly implemented because of the internal win-lose atmosphere that is created. E. H. Schein points out that:

> [A person advocating the minority position] often feels that the voting has created two camps within the group, that these camps are now in win-lose competition, that his camp lost the first round but it is just a matter of time until it can regroup, pick up some support, and win the next time a vote comes up. In other words, voting creates coalitions, and the preoccupation of the losing coalition is not how to implement what the majority wants but how to win the next battle.[2]

Fourth, a portion of the group's members may feel dissatisfied with a decision even though they participated heavily in the discussion. While their participation might have been high, their discontent stems from lack of influence. N.R.F. Maier comments:

> The research results clearly show that each participant's satisfaction with group discussions is related to whether she

had the opportunity to influence the outcome as much as she wished. Thus satisfaction depends more on a person's felt influence than on how much she talked at the meeting. Frequently members feel that if they had had more time, they could have changed the outcome.[3]

•

For voting to have any chance of being a constructive process, all members must be given time to understand and assess the situation, as well as provided an opportunity to exert real influence on it. If this is done, the minority coalition will have a far greater sense of obligation to go along with the majority decision.

In general, the vote is an appropriate and necessary device when you have a large group (any size larger than a dozen people) deciding an issue. Voting is rarely appropriate in smaller face-to-face groups, where consensus can be used as a much better alternative by avoiding the drawbacks of voting highlighted above.

Incidentally, the fourth point (on the role of personal influence) applies just as forcefully to the consultative autocratic option (CA II) discussed earlier as well as to both whole group options discussed below—consensus (WG I) and unanimity of consent (WG II)—as it does to majority vote. If real personal influence is not at the core of group processes, why bother?

Whole Group I: Consensus

Exercising the consensus option lets you share the problem with the group first and then facilitate the group members in processing information to *reach a solution each member can either fully agree with or, at least, agree to support.*

Three points are critical relative to facilitating consensus:

1. Keep pushing teammates to search for ways to have this *and* that—to do it both faster *and* cheaper, to have it be both simple *and* well-featured, to have it be both "green" *and* fully functional.
2. Avoid the "us versus them" polarization that often occurs, by continuously reminding all teammates that the focus is on defeating the problem rather than defeating each other.

3. *Verbally test for consensus* by going around the table asking each teammate in turn to state either agreement, support, or lack of support at this time for the solution in question. If every member declares agreement or support, you have consensus; if one or more teammates states that he or she can do neither, more processing must be done.

A vivid example of achieving consensus despite the issue's difficulty was the Presidential Commission, headed by former Secretary of State William Rogers, that investigated the space shuttle tragedy of January 28, 1986. Among other things, this 12-member commission uncovered many issues relative to NASA's role in the *Challenger* explosion. In writing their final report—a grinding task—11 members of the commission were reluctant to assign personal blame to specific NASA officials even though they were quite willing to strongly denounce the agency's "flawed" process. This alarmed the twelfth member, a physicist; he pushed hard for stronger language. He wanted to call some of NASA's managers "stupid," something the other members would not support, and the group stalled. After more dialogue, the commission resolved the matter by reaching consensus by allowing the dissenting member to record his own views in an appendix.

The commission had developed a win-win outcome. The physicist got his views into the report, but they could not be interpreted by readers as representing the view of the entire commission.

Because of its role in collaborative leadership and the positive impact it has in building team power within and across groups, consensus and its ramifications will be treated in greater detail in Chapter Six.

Whole Group II: Unanimity of Consent

This is the ultimate in decision making. To achieve unanimity of consent, *all group members must agree with a given course of action.* The strongest example of unanimity of consent is the Sixth Amendment to the United States Constitution, mandating unanimity in a criminal jury trial. For the vast majority of situations concerning public and private sector operations on a day-in, day-out basis, consensus will be more than adequate for making whole group decisions that everyone is committed to implementing.

On occasion when seeking consensus, a group actually winds up with unanimity of consent. This occurs when the manager discovers in going around the table, testing for consensus, that all group members agree with the proposed course of action, as opposed to having some state agreement and others state support. Unanimity as a by-product of consensus is fine, but because of the length of time it typically takes to achieve unanimity, it should be reserved only for the group's most critical decisions.

FORCES IMPACTING THE CHOICE OF A DECISION OPTION: KEY TRADE-OFF CONSIDERATIONS

Which decision option is best? As mentioned earlier, each has its time and place; they all can produce positive as well as negative consequences for group operations. What may be a well-timed and effective decision-making choice in one situation may, under different conditions, be an ill-timed and ineffective option. While any number of factors can affect a decision method's success or failure, Figure 5-2 highlights five that are especially pervasive and influential.

The center oval shows the deck of eight decision options available to you as a leader. These were covered in Figure 5-1. Surrounding the center oval are the five key considerations that have a bearing on which final option you select. In some instances, after applying the considerations, several options may turn out to be equally practical. In other cases you might have a preference, or believe you already know which option is best; however, after applying the selection factors, you may find that a different option is really the most appropriate one.

Figure 5-2 and its attendant discussion is valuable because it presses you to think through five critical dynamics in decision making that research has shown make a difference when it comes to implementation.

We'll cover all five forces, starting with *quality* and *acceptance* being treated together. In tandem, these two require some care-

Figure 5-2. Forces Impacting the Choice of a Decision Option: Trade-off Considerations

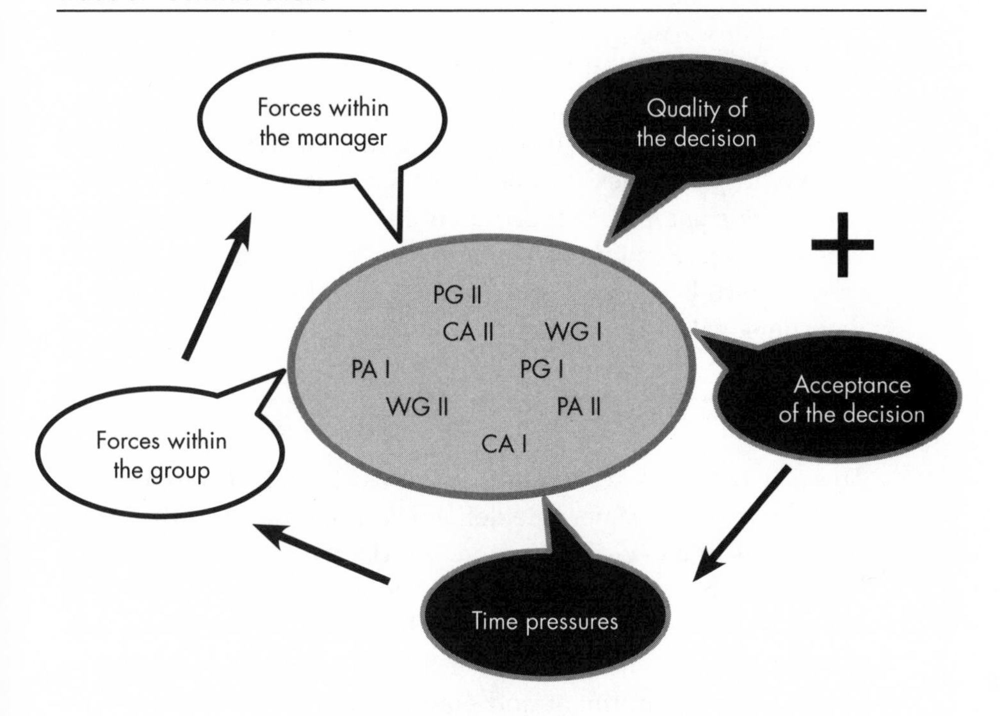

Your Trade-Off Considerations

Decision making is always a weighing and balancing of trade-offs, especially among these five dimensions.

Quality: The objective fitness of a decision; the purely objective or impersonal considerations that indicate one solution is clearly superior to another one.

Acceptance: The commitment and emotional support of those who must execute the decision (i.e., those who must make it happen).

Time: The time frame within which a final decision must be rendered (i.e., what is the deadline).

Forces: Forces within the team that have an impact on your final choice of a decision-making option.

Mgr. Forces: Forces within yourself, as manager, that impact your final choice of a decision-making option.

ful thinking regarding the requirements of the decision situation. We'll then move clockwise around the oval, taking *time pressures* next. Along with quality and acceptance, time pressure is the other prime consideration. These "big three" are shown in black to highlight them. Once you have examined these first three forces and zeroed in on an option, you will need to take two intervening variables into account, *forces within the group* and *forces within yourself*, to ascertain if the original option remains a workable one. Sometimes you will have to move to a more autocratic choice because of conditions within the group or due to your personal comfort level.

Decision Quality and Decision Acceptance Requirements

These two factors represent a framework first developed and studied by N.R.F. Maier.[4] He defined *decision quality* as the objective fitness of a decision; in other words, the purely objective or impersonal attributes that indicate one solution is superior to an alternative solution. For example, the payback of machine X is 10.5 months less than machine Y. Maier defined *decision acceptance* as the commitment and emotional support of those who must execute the decision. With acceptance, individuals and groups are willing to put in extra time and effort to make the decision work. They believe in what needs to be done, and won't settle for less than achievement of the objective.

Regarding quality and acceptance, Maier studied the four possible combinations of these two dimensions using an either/or classification. Let's now review what decision options are best aligned with each of these quality-acceptance combinations.

Neither quality nor acceptance is important. This type of decision is made from among alternatives that are equally good from a quality point of view; furthermore, the final choice makes little difference to those who have to execute it.

Any of the eight decision options in this chapter would be appropriate in this instance. However, because the pure autocratic methods—one-way pronouncement (PA I) and human filing cabinet (PA II)—would be the least time consuming, either could be the most favored. In the simplest situations, one could even flip a coin.

An example of a decision event in which quality and acceptance are both unimportant might be deciding where to hold a meeting, given a choice of three conference rooms equally comfortable and convenient for all attendees. Or what shade of gray to paint the factory floor.

Quality is important, acceptance is not. Such decisions do not require emotional commitment from the people who must execute them. Decisions about where the company buys its raw materials, the rate at which to expand facilities, or how much the firm charges for its products, are all examples. The success of decisions like these depends on careful analysis of the facts, figures, pros, cons, trade-offs, and the like among experts; a quality decision by the proper authority; then appropriate execution by skilled line and staff personnel.

In a situation where the quality of the decision is the main ingredient, and the requirement for acceptance is relatively low, the final decision has to rest with the person responsible for the activity. However, since other individuals may have specialized knowledge and technical information critical to making a high-quality decision, the person in charge must be capable of drawing upon this expertise by consulting with the specialists either individually or in a group before rendering a decision. Hence the consultative autocratic approaches (CA I and/or CA II) would be the most suitable decision options.

Acceptance is important, quality is not. This condition exists when there are a variety of possible outcomes with insignificant differences in quality. In this case, winning the emotional support of all people affected by the outcome is paramount. A few examples would be: deciding which three people in a group of six should work overtime; selecting who, from a staff of nine, gets to go to the national trade show; determining which three administrators from a team of ten will receive the new personal computers. Often these decisions *involve* a good deal of *emotion* as well as the issue of fairness.

This issue of fairness is a somewhat nebulous one—we don't learn much from the saying, "What's fair is fair." I've used five guidelines with managerial groups to get them through some sticky encounters where a scarce and highly desired resource—budget money, headcount, space, office location, perks—had to be apportioned among a group of people. The approach used was not to get the group members to agree on the allocation results per se, but rather to get them to agree on a fair method for making the allocation.

A fair method is any structured process that guides the allocation of scarce resources in a manner that assures the commitment and emotional support to the final outcome by all involved. To create a fair process, you and all other group members need to collaborate in making certain your process meets these five guidelines:

1. It shows no evidence of favoritism.
2. It does not permit the promotion of self-interest.
3. It has some mechanism that allows the process to be reviewed and modified, whenever and wherever needed, to keep it fair.
4. It is morally, ethically, and legally correct.
5. It allows all people affected by the process to have their concerns represented during its creation.

Since these high-acceptance situations have no best or correct solution, the fair, effective choice is the one most desirable to the people involved. Use of the whole group decision options of consensus (WG I), or in an extreme case, unanimity of consent (WG 11), would be the most fitting here. Alternatively, an open group discussion followed by a majority vote (PG II) might be advantageous, depending on the situation and the size of the group.

Quality and acceptance are both important. Examples include the redistribution of workload among the members of a staff following a workforce reduction, a change in the way products are designed (e.g., going from drafting boards to computer-aided design), or the institution of Total Quality in an organization. Maier states: "These problems frequently involve: (1) conflicts in interests (real or apparent) between managers and workers; (2) use of expert opinion; and (3) complex patterns of variables that create issues of fairness."[5]

In striving to optimize both the quality and acceptance dimensions, you will face your most difficult collaborative leadership challenge. The whole-group decision options of consensus (WG I) or unanimity of consent (WG II) are the ones required here. In your leadership role helping the group reach a consensus, or in the rare instance of unanimity of consent, you are attempting to do two things: (1) draw on the group's knowledge, skills, and expertise to optimize the decision's quality; while simultaneously (2) gaining the group's full commitment to the final outcome. This is no small task; still, learning and practicing the skills taught throughout this book—especially those covered in the consensus, conflict management, and problem-solving chapters—will ease your mission.

Time Pressure

Time pressure is one of the greatest stresses that any manager feels. As the vertical arrow running along the right side of Figure 5-1 shows, when urgency figures in decision making, the decision process is likely to shift from a more time-consuming option to a less time-consuming one. For example, time pressures may force a shift away from the more time-consuming consensus process (WG I) to the less time-consuming majority voting (PG II) "carrying the day." Time pressures may cause you to change from acting as a consultative autocrat processing information with the group (CA II) to being a pure autocrat using people only as human filing cabinets (PA II).

Still, selecting a decision option solely on the basis of expeditiousness overlooks two other crucial factors: (1) the time required for those affected to develop an understanding of, and commitment to, the decision; and (2) the time to execute the decision once it's made.

Obviously, as the unit leader, you can make decisions much more quickly in the autocratic mode than in the whole group mode. When employing either of the two whole group approaches, you are actually building commitment along the road to the final decision. On the other hand, concurrent commitment-building most likely is not taking place if you select the autocratic choices. Once you make an autocratic decision, often additional time must be spent persuading the implementers to "get on board" if commitment is essential. Thereafter, more time must be spent checking to ensure proper implementation.

A consensus decision approach, where you share authority with all group members, may often be more efficient than a purely leader-determined decision. The reason is, both the time required to reach a decision or solution in a problem-solving group, and the time needed to implement the decision, depend on the part played by group members in the decision-making process. Even though the shared leadership approach may increase the time initially required to arrive at a decision, it drastically reduces the time needed to develop the degree of commitment that ensures effective and rapid implementation.

The relationship between time pressure and the choice of a decision option is rooted in urgency. If the immediate need for a decision is driving you, you are very likely to decide to use one of the four autocratic options or the majority vote (PG II), despite the fact that it may generate additional time pressures when the decision needs to be carried out. On the flipside, due to the need for acceptance, you and your group may decide to go for consensus (WG I) despite time pressures. In doing so, however, everyone realizes that if consensus cannot be achieved, you will have to resort to a majority vote (PG II) or fall back to a consultative autocratic mode (CA II) to obtain a decision within the time constraints.

Forces within the Group

Thinking about the group's composition will influence the selection of a decision-making option. The maturity of the group members and their ability to work together are key factors permitting more group-oriented decision options. Veteran units tend to be superior in their ability to process information accurately and to handle conflicts among members. Newly formed groups spend too much energy avoiding controversy, worrying about what the boss might think, and being concerned with their own position and role in the team.

Forces within the Manager

Finally, you need to factor yourself—as a person—into the equation as you go about opting for a decision choice. You must take into account your technical knowledge and understanding of the issue in combination with your assumptions about people, the relationship of your personal style and managerial practices, and your beliefs about the appropriate use of managerial power. These personal aspects are all part of the mix influencing how a decision might be made in a given situation.

Table 5-1 summarizes the typical forces within the group and manager that impact the final choice of a decision option. This listing is not exhaustive, but it does highlight the primary tradeoff considerations. Think about your own leadership environment/situation and add other points to either side that are relevant for you.

Table 5-1. Group and Managerial Forces to Consider When Selecting a Decision Option

Forces within the *Group* to Consider When Choosing a Decision Option	**Forces within *Yourself* (as Manager) to Consider When Choosing a Decision Option**
• Level of knowledge, skills, abilities among teammates to understand the issues, to collaborate, and to make the decision. • Degree of teammates' interest in making the decision.	• Amount of knowledge and data you possess to make a quality decision that would be acceptable to all concerned. • Degree to which group development is a key factor. • Your general stylistic inclinations (autocratic vs. shared).

Table 5-1. Group and Managerial Forces to Consider When Selecting a Decision Option (continued)

Forces within the *Group* to Consider When Choosing a Decision Option	Forces within *Yourself* (as Manager) to Consider When Choosing a Decision Option
• Scope of past history in successfully working collaboratively to solve problems. • Extent to which politics and self-interests are present within the team. • Level of understanding and passion for what is trying to be accomplished. • Level of trust among teammates. • Degree of maturity of teammates in terms of understanding the organization, navigating the political system, and working with people in critical functions upon which the team depends. • Extent to which teammates' needs/goals align with task's goals.	• Extent to which boundaries are externally imposed on you so the outcome is pretty much a foregone conclusion. • Degree to which you believe the group is or is not qualified to make a final decision. • Level of your need for control, predictability, and stability—particularly when conditions are unstable or in flux. • Your level of trusting others. • Level of your ability to delegate and "let go" when stress and pressure mount.

PULLING IT ALL TOGETHER: GENERALIZATIONS ABOUT CHOOSING A DECISION OPTION

As pointed out in the previous section, there are no hard and fast rules to be followed that will "light up the sky" with a message stating unequivocally: "Given these circumstances, this is the decision style to employ." The sheer number of variables involved and their complex interactions makes this too difficult.[6]

However, by identifying and defining each situation in terms of the five major forces that both research studies and experience deem crucial, you will be able to acquire keen insight into the appropriateness of electing certain decision approaches over others. Relative to the ideas developed in this chapter, several implications stand out.

First, you have a deck of eight decision-making cards from which to select a decision option. This provides a great deal of latitude when making your choice.

Second, in order to improve the chances of selecting an appropriate option, you need to understand the role of the five major forces and how each affects the way your hand is played. As treated here, the major forces you need to consider are: the quality and acceptance ramifications of the decision, the inherent time pressures, the forces residing within the group, and personal forces within yourself.

For example, if the group members have useful information, if their acceptance is critical, if you do not have all of the knowledge and expertise to make a high-quality decision alone, if you feel secure facilitating a group decision, and if time pressures are moderate, then utilizing the whole group option of consensus would be appropriate. If the opposite conditions exist, then one of the autocratic approaches would be effective.

Third, most people do not want—or need—to be involved in every decision, especially those toward which they are indifferent. When they do contribute, however, they want those contributions to be taken seriously.

Fourth, trade-offs among aspects of the five forces will be the rule, not the exception. Consider a situation where the group members have useful information, their acceptance of the final decision is critical, and you do not have all the knowledge and expertise to make a high-quality decision alone. Now add your own insecurity in facilitating a group decision, as well as severe time pressures, and a trade-off must be made. You will have to decide if the time pressures and your own insecurities outweigh the other aspects of the decision situation. If they do, choosing one of the consultative autocratic options (CA I or CA II) could be an appropriate way to go; if they do not, selecting the partial group option of majority vote (PG II) may be a workable option. Moreover, consensus (WG I) might be chosen in spite of your insecurities and attendant time pressures because group acceptance is paramount.

Fifth, being flexible and using a variety of decision options is vital in successful decision making. The flexible leader is one who maintains a high batting average in accurately assessing the forces impacting the choice of a decision option and, based on the results of the assessment, is willing to make a sincere effort to behave accordingly. Because of circumstances, you may have to go with an option you are less comfortable with than some other alternative. In those cases, the experience you gain in working with the less-favored option will outweigh your feelings of discomfort.

DECISION MAKING: KEY LEARNING POINTS AND WHAT I WANT TO DO DIFFERENTLY TO IMPROVE MY SKILLS

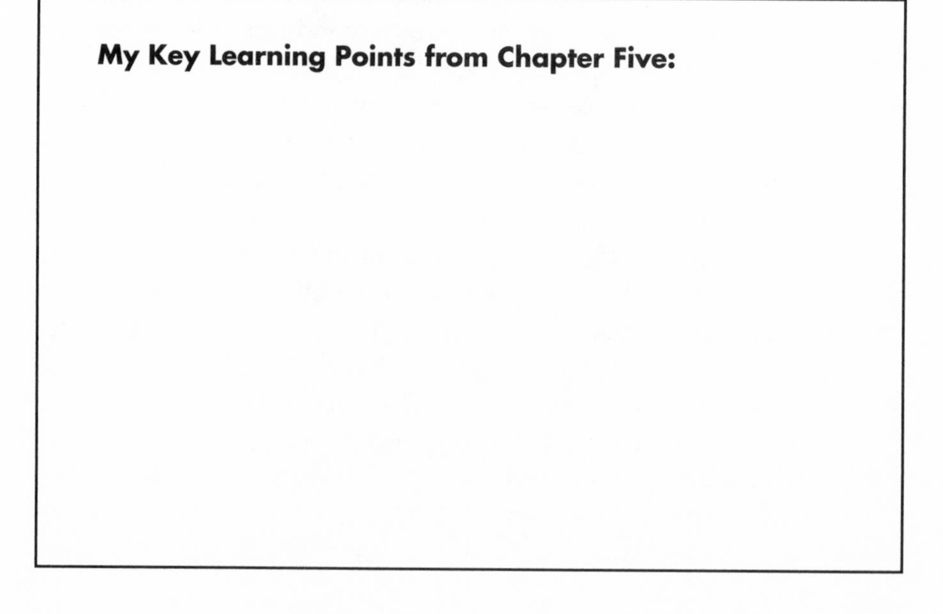

My Key Learning Points from Chapter Five:

What I Want to Do Differently:

READER REFLECTIONS AND APPLICATION ACTIVITIES FOR CHAPTER FIVE

Reflections

Every decision you make or facilitate, regardless of the option used, always carries the weight of legal, moral, and ethical imperatives. Use the *five gut checks* provided by Baldwin, Bommer, and Rubin as a reflection standard for all your decisions.

1. **The Wall Street Journal Test.** Would I stand by the decision if it made the front page of the *Wall Street Journal*? Would I be embarrassed if others knew I made (or facilitated) this decision?
2. **Platinum Rule Test.** Am I treating others the way they would want to be treated?
3. **Mom Test.** Would I be proud to tell my mother of this decision?
4. **Personal Gain Test.** Is the opportunity to gain personally standing in the way of my thinking? Have I given my personal gain too much weight?
5. **Cost Benefit Test.** Does this decision benefit some to the detriment of others? Have I considered its true impact on others?[7]

Leader Application Activity: Decision Making with an Individual (30 to 45 minutes)

1. Have teammate read Chapter Five and come prepared with answers to two questions in advance of meeting with you:

   - What did you feel were the most important learning points in the chapter?
   - Why were these key learning points for you?

2. Open the session by letting the teammate share his or her views on the two prework questions above. Actively listen, understand.
3. Share what you found important in the chapter and why these points were key for you.
4. Have the person present a real work situation to you and share the actual decision option used.
5. Mutually analyze the five forces that were in play (quality, acceptance, time, group, and managerial) at the time. Does your analysis indicate another option might have been more appropriate? Repeat the same process for two or three other decisions.
6. Now have the teammate present a decision he or she is facing. Mutually analyze it in terms of the five forces and agree on an option.
7. Do postdecision follow-up. How did the chosen option work out? Should another one have been used? Was the analytical process helpful? Why or why not?

Leader Application Activity: Decision Making with Your Team (45 minutes)

1. Have all teammates read Chapter Five and come prepared with answers to two questions in advance of the meeting:

   - What did you feel were the most important learning points from the chapter?
   - Why were these key learning points for you?

2. Open the session by asking teammates to share their views on the two prework questions. Actively listen, understand.

3. Share what you found important in the chapter and why these points were key for you.
4. Discuss how better decision making could be applied to improve our team sessions.
5. Which option(s) seem to dominate the decision-making process within our team? Why is that? Review a few recent decisions in terms of the five forces and discuss whether a different approach would have been more appropriate in light of the actual outcomes.
6. How often do we examine our decisions using something like the "five gut check factors" outlined in the Reflections piece? Is that something we should use routinely?
7. Before entering into major decisions, would our process be helped by discussing and understanding which option is to be used and why?
8. Can we get a consensus to apply the five forces and the "gut check" standards at every appropriate decision-making opportunity within this team and with other teams we individually lead?

NOTES

1. R. K. Mosvick and R. B. Nelson, *We've Got to Start Meeting Like This* (Glenview, Illinois: Scott, Foresman and Company, 1987), 49–50.
2. E. H. Schein, *Process Consultation,* vol. 1, *Its Role in Organizational Development,* 2nd ed. (Reading, Massachusetts: Addison-Wesley Publishing Company, 1988), 72–73.
3. N. R. F. Maier, *Problem-Solving Discussions and Conferences* (New York: McGraw-Hill Book Co., 1963), 146–147.
4. Ibid., 1–19.
5. N. R. F. Maier, *Psychology in Industrial Organizations,* 5th ed. (Boston: Houghton Mifflin Co., 1982), 172.
6. See V. Vroom and P. W. Yetton, *Leadership and Decision Making* (Pittsburgh: University of Pittsburgh Press, 1973). Vroom and Yetton have created a formal procedure that prescribes which decision styles best suit what situations. They developed a simple decision tree limited to seven yes/no answers to questions the manager asks to define the decision problem. With a specific decision defined,

the model indicates the appropriate level of participation required to make a quality decision. While far from being all inclusive, as well as being somewhat rigid in application, the model is a useful mechanism for bringing into focus the amount of employee participation, given the variables of a particular situation.

7. T. Baldwin, W. Bommer, R. Rubin, *Developing Management Skills: What Great Managers Know and Do.* (New York: McGraw-Hill/Irwin, 2008), 120.

CHAPTER SIX

CONSENSUS BUILDING

Facilitating Whole-Group Support

CHAPTER OBJECTIVES

- To remove widespread misunderstandings regarding consensus building
- To provide a process model for consensus decision making
- To detail six proven principles to be followed in any consensus decision-making effort
- To provide a set of eight tips and techniques for facilitating a consensus
- To provide a set of six alternatives for moving ahead when consensus gets stuck

INTRODUCTION

Two days before a famous beauty pageant was to be held, the well-known host, who also was an outspoken animal rights activist, discovered that the backdrop for the swimsuit competition would be a ski resort and that each of the semifinalists was initially to appear on stage wearing a fur coat. The host felt it would be hypocritical for him to be directly involved in an event that so blatantly violated his animal rights concerns. He threatened to quit his hosting job if the semifinalists wore fur coats. The show's sponsors said the elaborate and expensive scene was locked-in, with insufficient time to rework it into something else. Things were deadlocked and the whole pageant was in jeopardy at the eleventh hour. Getting a new host at that point to learn the complicated details of the show was impossible. After a day of intense discussion, the host and the show's sponsors reached a consensus to garb the contestants in synthetic pelts—something the host was quick to point out to viewers at the beginning of the swimsuit competition.

This example illustrates the underlying theme of consensus—various stakeholders working together to form a win-win solution out of a difficult set of circumstances. The win-win solution did not compromise the host's strong animal rights convictions, and at the same time it met the sponsor's requirement to keep the expensive and elaborate scene intact, saving their sunk costs and the contestants' past rehearsal time.

LINKING BACK TO THE PREVIOUS CHAPTER

The information in this chapter greatly expands the consensus material presented in Chapter Five. As you will recall, consensus (WG I) was one of the two whole group options from our tree diagram; unanimity of consent (WG II) was the other whole group option. Because of its impact on group processes and its role as a primary decision option, consensus building is given more extensive treatment here.

With respect to consensus versus unanimity of consent, while unanimity is rarely required as a decision option, the approach to generating it is the same as for building consensus. Before proceeding with a complete explanation of consensus, please remain mindful of the following points. From the previous chapter, we know that eight options are available when we want to make a decision.

- Consensus (Whole Group I) is one of them.
- Consensus is not appropriate for every situation and every decision; it should never be considered the only way or the best way to make decisions.
- Exercising the consensus option lets you share the problem with the group first and then facilitate the group members in processing information to reach a solution each member can either fully agree with or, at least, agree to support.
- Used inappropriately, consensus is a frustrating and time-consuming process that can create hostile, unyielding, and polarized cliques that prevent the group from reaching any decision.
- Used appropriately, consensus is a vibrant and energizing process that builds joint ownership, enthusiasm, and unified commitment to action.

CONSENSUS IN PERSPECTIVE

Consensus is derived from *sensus*, meaning a mental process, not from *census*, meaning counting. Thus it refers to minds coming together. Consensus is one of the most powerful skills for building team power. A leader and teammates working hand in hand to form a consensus around a course of action will find—regardless what the final decision turns out to be—that the result will be fully supported. What's more, all the people involved in the consensus building process will implement the final decision with determination because the team, together, has examined each facet of the problem and through debate, discussion, and fruitful friction unearthed the best way to proceed given the circumstances.

The real key to making the consensus process work is for all parties, leader and teammates alike, to describe how they see the issues. This exchanging of viewpoints is an opportunity for team members to learn from each other. Effective team members aim to remain open-minded, nondefensive, and flexible, rather than pushing for their self-interests at the expense of the team's greater good. When individualism starts trumping the needs of the team or the other stakeholders, consensus invariably fails.

The leader also has to ensure that team members have the time to bring forth the open dialogue necessary to share perceptions about various decision alternatives taking hold and what each one means to them. Only after these initial viewpoints are clear can the team proceed to highlight areas of agreement, areas of disagreement, and collaborate on a solution everyone can fully support. Leadership is critical in the consensus process, and no one captured the essence of this point better than Martin Luther King when he said, "A genuine leader is not a searcher for consensus but a molder of consensus."

Consensus is the means for transforming group knowledge and energy into effective action. Mutual trust, shared goals, complementary skills, accountability, character, commitment, and authentic communications, all covered in previous chapters, apply directly to consensus building.

As emphasized in those chapters, many aspects of the different collaborative leadership behaviors intertwine, and some will unequivocally affect other collaborative behaviors. As we detail the processes and techniques for consensus building, be alert to the connection between what you've learned previously. For starters, here are four points to keep in mind.

1. *Consensus is learning in action.* Team members must be committed to the idea that every consensus decision session is a learning opportunity. Everything that goes on in the attempt to mold a consensus is "grist for the mill." There are no failures—only learning opportunities and an enrichment of collaborative relationships.
2. *Consensus must operate in a trust-based environment.* Consensus rides on a road of trust. Consensus is built by people communicating openly and honestly with each other, by showing respect for each other; it is built by people being able to do what they claim is their ability to do and by people doing what they promise to do. Consensus building, like trust building, requires all parties involved to collaborate

rather than compete, judge, or blame. It is critical that once a consensus is reached, all team members follow through on their commitments, and reopen discussions if something problematic occurs during implementation.

3. *Consensus building means team members must value differences.* There will be a lot of disparity and diversity during the search for win-win solutions. Everyone in attendance needs to encourage and honor differences because these differences will be the wellspring of creative proposals, new ideas, and critical thinking.
4. *Consensus is "we thinking" vs. "I thinking."* By viewing each other as a team of collaborative partners as opposed to a collection of self-serving individuals, a consensus for the greater good of the team can be achieved.

A PROCESS MODEL FOR CONSENSUS DECISION MAKING

Building consensus is a flow of activities, and the model in Figure 6-1 portrays the primary activities and how they fit together. "Stage setting" is a necessary first step to get the problem or issue properly framed by the leader (or other person bringing up the subject for processing). Also, it is during stage setting that a backup option is declared so all team members are clear on the decision option to be used in case consensus cannot be achieved. This avoids any last minute surprises and charges of foul play.

Then, in steps two and three, the team collaborates in sorting through its information, data, and offered proposals to create a set of concrete potential solutions; this set is processed further, and narrowed down. At an appropriate point, as indicated by step four, the leader calls for an explicit test of consensus around a particular alternative. If consensus is achieved, then action plans for implementing the decision solution are created.

If consensus is not reached when tested the first time, then there's a second go-round, where the group members—as collaborative partners—solicit new concerns or objections, revisit old ones, and generate new proposals and ideas among all members

Figure 6-1. Process Model for the Consensus Decision-Making Option

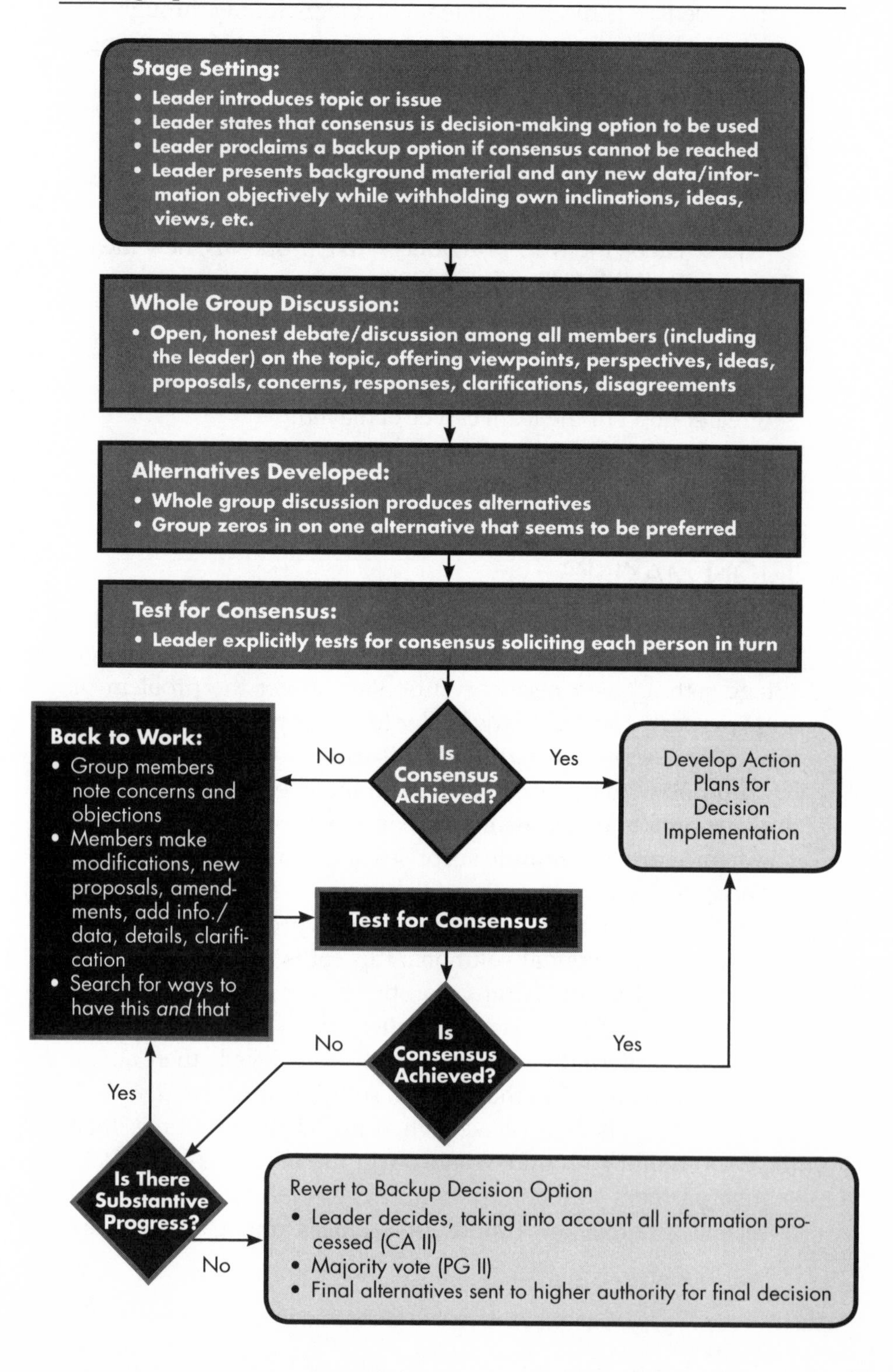

in an attempt to find some alternatives around which everyone can rally. At an appropriate juncture, the leader again calls for a test of consensus. If this second round of processing produces a consensus, the team then shifts into the action planning phase.

Should consensus still not be reached, the leader and the team have to assess if substantive progress is being made in an effort to reach a consensus. If yes, then a third (or even a fourth) round of processing may be needed before all facets of the issue can be ironed out and a consensus is molded. When that happens, then it's on to action planning.

If after several tests for consensus, people are dug in and substantive progress toward joint action is stalled—especially with a fixed deadline looming—the leader can call an end to the proceedings and use the backup option to get a decision completed.

SIX PRINCIPLES FOR CONSENSUS DECISION MAKING

For consensus decision making to work effectively, there are six proven principles you must adhere to anytime you decide to use consensus in a particular situation. Violate these principles at your own risk.

Always Think of Consensus as Win-Win, Not Compromise

The essence of consensus is win-win. In a decision-making situation, win-win indicates the development of a solution that does not dilute any strong convictions or essential needs of individual group members. Consensus, then, is a cooperative effort to find a sound solution acceptable to everyone, rather than a competitive struggle in which an unacceptable solution is forced on others.

Compromise, on the other hand, is a solution that "splits the difference," "strikes an average," or "meets everyone halfway." The problem with a compromise is that the outcome doesn't meet the needs or expectations of everyone involved. Compromise implies "halfhearted" agreement; it is a partial solution that often leaves group members unenthusiastic. There is doubt, lingering disagreement, and the potential for second-guessing the decision. In fact, there are always some losers and winners in a compromise decision. When that happens, the losers may be uninspired about putting that decision into action, and the winners become disenchanted with the efforts the losers are making. In some ways, a compromise can plant the seeds for later conflict. The episode in the sidebar portrays the contrast between win-win and compromise.

Compromise decisions also can produce some fractured results, as exemplified by a church that had half pews and half seats in the sanctuary because the renovation committee couldn't come to a consensus on which way to go!

Janice and Lenny are having breakfast. A bowl of fruit containing one orange is on the kitchen table. They both reach for it at the same time. Janice says, "Well, I can see we both want that orange." Lenny says yes and they argue over it for a minute. They decide to end their tiff by splitting it. Janice cuts the orange in half and gives Lenny his piece. Lenny peels his half, eats the fruit, and puts the peel down the garbage disposal. Janice peels her half, throws the fruit down the disposal, and keeps her half of the peel to be grated into the batter of her famous cake she plans to bake for a friend that afternoon.

The outcome of this interaction was a compromise. Each person's needs were only partly met. Instead of settling for 50 percent of what each desired—Lenny only getting half an orange to eat, which just partially satisfied him, and Janice only getting half the peel required for her cake batter, causing her to complain later that the cake will be mediocre at best—both could have had their needs fully met. Had they spent more time understanding each other's needs, they could have developed a win-win solution, 100 percent of the fruit for Lenny and 100 percent of the peel for Janice.

Win-win alternatives seldom are obvious to the group members attempting to decide a course of action. In striving to build a win-win consensus within your team, staying true to the integrity of the process will serve you well. Keep digging and pushing members to elaborate on what they would add, subtract, reconfigure, or otherwise modify in the alternatives being considered so revisions may be made to earn everyone's support. Include everyone in proposing revisions and building on them. Ask, over and over, "What do we need to do to have both this *and* that?"

Individuals who do not agree with the most preferred alternative quite often will join with those favoring it and mold a firm consensus, if three integrity conditions have been met: (1) they have had sufficient opportunity to present their viewpoints to the team, (2) they are sure their ideas have been listened to and truly understood, and (3) they can see that the group considered their ideas in formulating the final decision.

The fire stations and the day care centers. The sidebar of consensus in action on the next page has been summarized from M. Doyle and D. Straus's *How to Make Meetings Work.* They described a city of 100,000 in the West, undergoing rejuvenation. Its population was growing and new construction was cropping up everywhere; however, this prosperity brought problems.

Bottom line, if a group can reach consensus, the effort will be worthwhile, since there will be no disgruntled minority refusing to support, or coerced into supporting, the group action. The search for consensus places a premium on trying to understand the other person or constituency rather than arguing or forcing them into submission.

Determine in Advance the Fallback Decision Option if Consensus Cannot Be Reached

Most decisions must be made by a deadline, and generating consensus does take time. Therefore, do not impose unrealistic time limits that doom the consensus effort before it begins. Group members will quickly see through this manipulation. If tasks repeatedly have time frames too short to allow the neces-

The city had accumulated a $3 million surplus, and two well-organized and vocal coalitions were vying for these funds. Coalition One included homeowners and members of the firefighters' union who were concerned that the city had only an AA fire insurance rating due to its aging fire equipment and facilities. If the city used the available surplus to modernize its equipment, it would be able to get an AAA rating, homeowners' insurance rates would decrease, and the city would attract additional federal funds for housing. Coalition Two, comprised of various women's groups, had been working hard to publicize the city's lack of adequate day care centers. The $3 million could be used to rectify this deficiency, they argued.

Both sides agreed there was not enough money to finance new day care centers *and* new fire stations and firefighting equipment. Also, three members of the city council were up for reelection at year-end and they knew that taking either position would meet strong opposition.

Facing this predicament, the city council urged both parties to work with representatives of the relevant city departments, state and federal agencies, and experts from the insurance companies. The two coalitions agreed to work with these other groups, and in doing so they developed a win-win solution.

Part of the $3 million was used to convert the five outdated fire stations into day care centers. The day care centers thus would be able to operate on matching funds from state and local agencies. The bulk of the $3 million city budget surplus was then used to construct three regional fire stations and supply each with the latest firefighting equipment, allowing the three stations to serve the city more efficiently than the old setup. This resulted in an AAA insurance rating.[1]

Both coalitions got a win. The city got new fire stations and new equipment without having to remodel the old buildings, homeowners were pleased with the AAA insurance rating, and day care advocates got inexpensive, well-located facilities and the funds to operate them.

sary debate, discussion, analysis, insight, and buy-in relative to issues and proposed solutions, the members will soon become frustrated with the process of consensus building and curtail its use as a legitimate decision-making option.

Still, there will be occasions when the time frames are realistic, when all members of the team will work diligently to preserve the integrity of the consensus process, when everyone will pour their hearts and souls into the search for a win-win alternative, and yet be unable to bring about a consensus. That is, the group will be unable to uncover a course of action that all members either can agree with or can support.

When a decision is due, there are only three legitimate options to get a decision made. As noted in Chapter Five, there is a fourth option of decision by default; that is, doing nothing—drop the effort or postpone the decision—and let nature take its course. While real, doing nothing is not considered a legitimate option here because it is decision abdication, decision avoidance. Let's highlight the three legitimate fallback options available when consensus cannot be reached.

1. The unit manager, or the person with final decision-making authority in other types of hierarchical groups, can shift to the consultative autocratic option by making the final decision, but doing so in light of all the sharing and processing of information that has transpired up to that point. (This moves the decision option from consensus to CA II).
2. A vote can be held, with the majority opinion determining the final outcome. (This shifts the option dynamic to PG II).
3. A majority and minority position paper can be written and the issue passed to a higher authority for resolution. (Here the team gives up control of the final decision).

However, it is important to point out that in situations involving various volunteer groups, boards, councils, committees, and task forces, the fallback position is limited to either the second or the third alternatives. In other words, in meetings where the group leader lacks final decision-making authority, the only available alternatives to getting a decision rendered—if consensus has failed—are majority vote or escalation of the decision to a higher authority. In councils, committees, and so on, the leader often lacks the option of halting the information processing and providing the decision, because the authority and responsibility for making the decision are fully shared with the other group members and cannot be taken back.

Employing the Consultative Autocratic Option. After striving to achieve consensus, but compelled by circumstances to shift to the CA II option, the most effective process for you to employ would be: (1) summarize the different positions and interests of the team members—to demonstrate your understanding of all the positions and interests being advocated, (2) take all of those positions and interests into account, (3) make the final decision, (4) announce it, and (5) report the rationale for the decision back to the whole team.

If a genuine effort by everyone fails to result in consensus, reverting to the consultative autocratic method should not be considered a facilitation flop. At least the group understands the issues, knows who stands where and why, and sees the difficulty of making the decision. Also, as a result of attempting to build a consensus, you have been the recipient of valuable information, new ideas, and maybe even a solution not apparent at the beginning of the dialogue.

Finally, research and experience have shown that group members tend to be more understanding and accepting of a decision made by the manager after their full and meaningful involvement. That is, most of the time, little is lost in terms of group commitment when the Consultative Autocratic II option is chosen after a genuine effort to facilitate consensus.

Employing the Majority Vote Option. Using the PG II option —majority vote—the team leader or the chairperson of a volunteer group, committee, council, task force, and the like, calls for a vote to arrive at the final decision. While majority vote is a simple method for obtaining a decision when time has run out for consensus, it is not the perfect alternative. Majority vote by its nature produces win-lose decisions. For example, in a group of 12 people unable to reach consensus—eight advocating position A and four backing position B—a vote merely produces a decision that was already obvious: position A *wins* over position B eight to four.

Voting can be a suitable strategic option when a group meets over time; members sometimes vote on the winning side and sometimes on the losing side, and the alternatives voted on are reasonably acceptable to all. However, if certain people continually lose, and if they view such losses as personal defeats, then majority rule to resolve time-constrained consensus decisions can eventually destroy the team.

Elevation to a higher authority. Sometimes a situation arises where, after an open, honest effort, consensus cannot be reached. The group leader does not have the authority to make an autocratic decision, and since it produces winners and losers, voting is not deemed to be an acceptable decision-making option. Under these conditions, the only way to get a final decision is to elevate it to a person who does have the requisite authority.

Routinely moving decision making up the chain of command is not a preferred method. It should be used only after every means to a win-win consensus solution has been exhausted. However, if elevation is the only recourse, both the majority and minority positions must be written clearly, concisely, and without attacks on each other, so the higher-level decision maker can most fairly assess the data and make the decision.

Anytime a group sets out to make a decision by consensus, the fallback option—CA II, majority vote, or elevation to a higher authority—must be determined before information processing begins. Once a contingency option is predetermined, team members are quite motivated to reach consensus in order to avoid resorting to a less preferred backup option. Also, since all team members know the fallback position in advance, surprise and resentment over how the final decision ultimately gets made is eliminated.

At Key Decision Points, Combat the Illusion of Consensus by Explicitly Testing for It

The *illusion of consensus* is the most common trap to snare the unwary facilitator/leader. The trap comes in two forms: the *silence trap* and the *hubbub trap.*

The silence trap. The purchasing team manager says, "Let's see how we stand. Does anyone have an issue with buying the part from Melcher Manufacturing?" (Silence from group members.) "Okay, it looks like we have consensus; I'll call Melcher tomorrow and place the order."

In this episode, the manager takes the group's silence to mean consent. If no one spoke up against the proposal, then he assumes that everyone must be for it. But group members who are not fully committed to the course of action often hold back their comments because they assume that everyone else's silence implies agreement or support, and they are reluctant to disrupt the "unity" of the group.

The hubbub trap. The store owner says, "Is everyone okay with the plan to open up an hour earlier on Tuesdays and Thursdays and to close two hours earlier on Sunday?"

Two group members remain silent while the other five erupt in simultaneous chatter: "Yes." "Great." ... "I love it" ... "Aw right" ... "Best decision we ever made."

The store owner then says, "Fine, I see we have an enthusiastic consensus. We can make the changes starting the first week of next month."

This situation is more subtle than the previous one. Here, two silent members have not indicated their position. Are they for or against the change-in-hours plan? The manager can't know for certain unless she asks them. What she does know is that there is a vocal groundswell of enthusiastic agreement from the group. Unfortunately, this hubbub of confirmation from the majority masks the silence of the other two people, and the manager, reinforced by the chorus of support, declares she has a solid consensus.

In both of these examples, the manager has fallen victim to the illusion of consensus. *Silence does not mean consent!* Maybe all of the silent members involved in each of the two situations were genuinely on board and committed to the decisions just made; on the other hand, maybe they were not. By accepting silence as consent or by missing a few silent members in the hubbub of affirmation, neither manager knows for certain if genuine consensus has been attained. They both have assumed it—a very dangerous course that could produce bad consequences down the road during or after the decision has been implemented.

Therefore, don't equate silence with consent. Find out what the silent ones are thinking. Quiet members are often harboring valuable ideas or insights that are contrary to the consensus being shaped. As a collaborative leader, the only way for you to *avoid* being seduced by the illusion of consensus is to *explicitly test for it.*

If each group member responds that he or she either agrees with or, at a minimum, agrees to support the alternative under consideration, a consensus has been cultivated and you can move the team on to action planning. If one or more members can neither agree with nor support the action in question, consensus has not been achieved and the information processing and search for acceptable solutions must continue.

> Explicitly testing for consensus means going around the table asking each person one by one the following question:
>
> "Do you *agree with* the proposed course of action, *or* if you do not agree with it, *will you support* the proposed course of action?"

In those situations when, after explicitly testing for it around the table, you find that consensus has not been attained, you loop back to more information digging and processing with the group. Start by asking each of the dissenting members to elaborate on what they would add, subtract, or otherwise modify in the alternative being considered so that changes might be made to win their support. (Review Figure 6-1 for the full process.) At no time assume a consensus—always test for it.

Develop Shared Values Regarding Consensus Decision Making

In order to put the consensus decision-making process in proper perspective and enhance its effectiveness, all task-oriented groups, whether permanent or temporary, need to take some time to decide what consensus means to them and how consensus will operate within their group. Consensus cannot be a win-win process unless all members agree on what it means and how it functions within their group.

As part of a three-day team-building session, I had the opportunity to facilitate a team comprising the president and the 11 senior vice presidents of Xerox's Business Products and Systems Group while they determined how consensus decision making would operate within their team. The discussion was intense, lasting several hours before the senior staff achieved unanimity of consent on their process as shown in this document:

This management team's approach to consensus is presented here because it is more than just the shared values of a high-level executive team. It is an outstanding definition of how consen-

Shared Values Regarding Consensus Decision Making

Whenever we, the Senior Executive Staff, have decided to make a decision by consensus,* and I am not able to agree with the course of action being advocated by the majority, I can state that "I will support" the preferred course of action (meaning I will do my very best to ensure its successful implementation) if my most essential needs were met and the integrity of the consensus process has been preserved. Preserving the integrity of the process means free and open debate and discussion of the issues were maintained and my concerns were fully understood and considered as the preferred course of action evolved. Stating "I will support" puts me wholly on board as an owner of the decision and makes me a part of the consensus. (*Note that consensus is only one form of decision making; this process only applies to consensus.)

sus decision making should operate on a consistent basis in any group. Each executive signed the document, and its principles became integrated into the senior staff's everyday way of working together.

Let's examine the power behind these words. The executives defined *support* to ensure that anytime one of them says "I will support" a particular course of action, that person is fully aware of the obligation he or she has to the staff. At the same time, all the other staff members are certain about what that obligation is. Also, the senior staff's shared values put a premium on maintaining the integrity of the process. While the final decision most likely will not reflect the exact wishes of every executive, *if* it embodies the *most* essential requirements of each member, and *if* everyone's concerns were understood and taken into account as the final decision was shaped, that decision still can be willingly supported by those executives not in complete agreement with it.

Andrew Grove, a founder and former CEO of Intel, in his book *High Output Management*, addresses the subject of "how can I support something I don't agree with?"

> Many people have trouble supporting a decision with which they do not agree, but that they need to do so is simply inevitable. Even when we all have the same facts and we all have the same interests of an organization in mind, we tend to have honest, strongly felt, real differences of opinion. No matter how much time we may spend trying to forge agreement, we just won't be able to get it on many issues. But an organization does not live by its members' agreeing with one another at all times about everything. It lives instead by people committing to support the decisions and moves of the business. All a manager can expect is that the commitment to support is honestly present, and this is something he can and must get from everyone.[2]

It pays to spend the necessary time to develop shared values around the issue of "How do we want consensus to operate within our work group?" Once this framework is in place, disruptive friction, misunderstandings, and hedging of commitments will be greatly reduced.

Stamp Out the Declaration "I Can Live with It"

In the role of process consultant working with many different work groups, I became keenly aware of a very subtle but crucial difference in the way groups practiced the consensus process. The difference was use or nonuse of the statement: "I can live with it."

Groups that routinely used and accepted the I-can-live-with-it pronouncement to signify support for a given proposal were observed to reach nominal consensus—consensus in name only. Their nominal consensus was hollow and lacked substance because the vital ingredient—genuine commitment to the final outcome by all group members—was lacking.

By verifying my observations with meeting participants in a variety of organizational settings, I made three profound discoveries:

1. The true message of the "I can live with it" declaration was ambiguous. At times it was used as a sincere statement of genuine support for the alternative being considered. More often than not, however, the phrase was used to give an *impression* of full support, while in reality the true message being communicated is: "I can live with the rest of you going off and implementing this decision, but don't count on me for much help in carrying it out because I'm not committed to what the rest of you want to do." Used inauthentically, the I-can-live-with-it statement can be a means for terminating the consensus process without really committing to the course of action being decided.
2. In a majority of cases, groups that make a decision based on nominal consensus were forced to confront a significantly greater degree of destructive conflict, uncooperativeness, finger pointing, and misunderstandings among group members during the implementation phase than were groups that worked out an honest consensus.
3. Groups that build a genuine consensus were able to do so in part because they virtually outlawed the "I-can-live-with-

> it" declaration from all of their consensus deliberations. The only acceptable affirmations for indicating consensus were "*I agree with* the course of action" or "*I will support* the course of action." Groups that were observed welding true consensus abided by a rule that was crystal clear: If a group member was unable to state agreement with, or support for, the alternative under consideration, that person was not considered on board, and therefore consensus had not been achieved. It was that simple.

The major implication of these findings is that "I can live with it" has no place in the consensus process. The statement is ambiguous and does not indicate a genuine buy-in to the proposal being deliberated. It produces a brittle consensus, giving the "I can live with it" people a back door through which to run the first time the team faces implementation difficulties or an organizational challenge, which it inevitably will.

Use the Consensus Option at Key Decision Points

Consensus should be used for *major elements* of a decision, not the many specifics. For instance, problem definition, decision assumptions, solution evaluation criteria, final solution choice, and priority setting are a few examples of major decision elements.

Since multidisciplinary teams are common in business, design, and manufacturing environments, it is often problematic for the new teammates to work effectively together due to technical vocabulary differences or misunderstandings over process terminology. This issue especially disturbs design projects. Design projects are composed of people with specific, diverse knowledge, from different disciplines, to execute special activities. Using consensus as a means to iron out and agree to common vocabulary, terminology, acronyms, procedural formats, and so on early in the process, will go far in creating valuable understanding and reducing conflicts in working relationships for all the teammates involved. The ultimate result of using consensus this way: better productivity.

EIGHT PRACTICAL TECHNIQUES FOR CONVERGING ON WIN-WIN CONSENSUS SOLUTIONS

The journey to consensus is rarely straight or smooth. Therefore, when your team has trouble coming together on a common decision solution, you will be able to keep the group moving toward consensus and away from tangents, bickering, and confusion by drawing from the following eight tips and techniques.

Encourage Flexibility: Exploding the Mind-set That It Has to Be All or Nothing.

As a collaborative leader, one of the most important techniques for advancing consensus is relentlessly encouraging flexibility. Proposed ideas and options are not always mutually exclusive, though at first they may appear to be. Proposals should be treated as a pool of information to be taken apart, examined, modified, extended, and reconfigured in the pursuit of an outcome that everyone can either agree with or agree to support. The power of flexibility in facilitating consensus is underscored by Doyle and Straus:

> Don't assume that you have to choose a single alternative. Sometimes several alternatives can work together. It does not always have to be either Plan A or Plan B, but both A *and* B: price cuts *and* salary reductions, high-rise *and* low-rise buildings, cleaning the house *and* going to the movies. Don't force your group to choose between two or more compatible and feasible alternatives. . . . Encourage people to surface and keep voicing their concerns until their fears have been dealt with. They are doing the group a valuable service by pointing out shortcomings of the solution; in the end the quality of the final solution will be higher. . . . Build on success: "Look at how many points we agree on." Encourage flexibility. Keep dancing around the problem. Try different approaches.[3]

Rather than viewing the decision process as a method that inevitably produces winners and losers, you must remind and encourage teammates always to look for ways to include parts or all of every good alternative into the final decision. If most of the team is homing in on alternative 2, you need to be alert to ask the question: "How can we incorporate some features from alternatives 1, 3, and 4 into our final decision as well?" If teammates see pieces of their own ideas in the final solution, they will be willing to support it even if it was not their preferred outcome. Also, combining ideas consistently improves overall decision quality.

Start with the Alternative That Has the Least Opposition: Generating Momentum When No Alternative Stands Out

Imagine you have been working through a complicated and emotional issue with your teammates for a couple hours; you've narrowed things down to a significant few possibilities. These remaining few were then openly examined, taken apart, moved around, refined, and combined into a newly reconfigured set of solutions in the search for a win-win outcome. Nevertheless, the team discovers it is stalled in its effort to achieve consensus. What to do? Try *negative voting* to identify the least opposed alternative.

Let's assume eight people, including you, are wrestling with three newly configured decision possibilities. Pointing to alternative 1, ask, "How many people *cannot support* (item 1 at this time?" Five hands go up including yours. You would write "five" next to alternative 1. Point to 2 and ask the same question; this time six hands go up not including yours. Write "six" next to alternative 2. For alternative 3, assume three hands are raised in opposition.

Because alternative 3 has the least current opposition, it becomes the base for building consensus. You would then concentrate on facilitating a full understanding of why the three dissenting members cannot support the third alternative. Record their comments on a flip chart. and once that has been done—as a whole team—process the information, search out new proposals, modify and build on ideas to make alterations that would satisfy the three not yet fully on board.

Use the Swiss Cheese Approach: Gaining Momentum When Differences Are Numerous and/or Complex

Consensus grows out of collaboration. As long as individuals or coalitions are attacking each other and talking about differences rather than attempting to collaborate, the possibility of forging consensus around a particular solution is virtually nonexistent. Group members often resist a joint search for options and the mutual creation of a win-win solution because the number of identified differences overwhelms them. Moreover, the complexity of the differences (politics, values, economics, timing, past history, etc.) may shatter the group members' confidence that consensus is possible. In this situation, the manager should bring the "Swiss cheese" technique into play.

The Swiss cheese technique suggests that the total problem often seems like a huge slab of cheese: overwhelming. However, by taking several bites and thereby putting a few holes in the slab, the problem does not seem so oppressive. Whenever the group resists making a collaborative attack on the identified issues, you must pick one issue that group members agree is important and focus everyone's attention and energy on this single problem: bite 1. In rallying the group, you should ask the members: (1) to be optimistic that this single problem can be resolved, and (2) not to get hung up on divisive, destructive arguments. Once this first collaborative bite is taken, the second and third collaborative bites can be taken in the same manner. Then, with three bites out, the group will have gained some momentum operating as a collaborative unit. The original matter will not seem so overwhelming now, plus a "can do" spirit will have been generated.

The sidebar on the next page provides an example of how I used the Swiss cheese method to great benefit in a sticky situation while working at Xerox.

Utilize a Committee from the Whole Group: Creating Consensus in a Small Group to Use as a Springboard to Consensus within the Whole Group

If a large group (say 10 or more) is bedeviled in its attempt to reach a consensus, break the large group into a small subcom-

I was asked to work with the senior teams of two archrival systems engineering units that needed to be merged into one unit. They were always fighting over the same resources; each believed it was superior to the other, and animosity was boiling over on both sides regarding the merger. Bringing the principals from both sides together to work out the myriad complexities seemed like an impossible task because the moats around both fiefdoms were wide and deep.

Instead of hitting the merger straight on, I asked the group members to be optimistic and assume all problems had been resolved except for the name of the new department. I worked that issue—and that issue only—until a consensus was achieved. With a lot of sailboat owners on both sides, after a while they enthusiastically reached consensus on the name Systems Engineering Applications (SEA).

Then I asked what factors would make SEA an outstanding unit with respect to the total organization. Once more the focus was strictly on this topic. They dug in—using a lot of sailing and sea metaphors—to come up with seven key operating principles, each beginning with the letter C (sailing the seven C's). Now, with these two bites taken and the ice broken, they began working as collaborative partners on the real issues and subsets of the merger they had so adamantly resisted before.

mittee—containing an equal number of people holding each primary position—and charge it with going off-line and working the issue. The typical assignment given to this kind of subcommittee is to take all the information generated by the whole team, draw from it, and achieve consensus on any amendments, modifications, resolutions, and/or new proposals. A deadline for reporting back its results to the whole team is also agreed to.

When presented, the subcommittee's new ideas can jump-start the rest of the team by offering fresh proposals or, better yet, possibly the solution that accommodates the essential needs of everyone. In any event, the whole team now has the opportunity to process, modify, and select from proposals that have already been thrashed through and given the stamp of consensus by the subcommittee.

Take Breaks: Calling Timeout to Reflect, to Regroup

Breaks are a useful tool. Calling for a break, for a few minutes or a day or two, until another session is convened, serves a number of purposes in the facilitation of consensus. It gives members time to be alone, to reflect, and to reconsider positions, interests, roles, and behaviors. When tempers begin to flare and positions start to harden, breaks allow people to cool off and become more rational. Breaks also give group members—including the leader—a chance to "work issues off-line." Members can meet in small groups and resolve differences among themselves. When a behind-the-scenes resolution is worked out and conveyed to the rest of the group, often a roadblock is eliminated and the full group is revitalized in its effort to form a win-win outcome.

Make Use of Straw Votes: Maintaining an Ongoing Sense of the Consensus

Trying to facilitate a consensus without being attuned to the ebb and flow of who stands where as the proposing, counterproposing, and building take place is facilitating blind. Periodic straw voting to maintain a sense of the consensus is critical because it highlights who does and who does not support a particular decision as currently formulated. Objections then can be sorted out so the group can unite in eliminating them.

Straw voting is an informal, nonbinding show of hands to test the number of people in a group who agree with or support a particular decision. A straw vote may be used to estimate how close the group is to consensus, and whether it is time to start struggling to finalize a decision or whether much more discussion is necessary. In larger groups, some people do not speak up. A straw vote is a way for silent people to express their opinions and feel that they are being given a chance to have input.

Should you decide to use a series of straw voting activities over the full course of deliberations to achieve consensus, the following message must be communicated loud and clear: The straw vote is not a movement to *majority* rule. It is merely a means for gauging how close the group is to consensus, and for identifying the most serious objections for further discussion.

Seek a Conditional Consensus if the Group Is Polarized: Getting Support to Try a Particular Option for a Certain Period of Time

Commonly, the consensus effort grinds to a halt with several options still open and members arguing their rationale for a particular option—but without any hard facts, figures, or past history to substantiate their arguments. When this occurs, shift the facilitation angle from shooting for full consensus to one of conditional consensus.

Conditional consensus is the willingness of team members to give a particular solution a trial run. Depending on the specifics, it may be possible to give several proposed solutions trial runs simultaneously. In any case, conditional consensus represents support to move forward, try out different options, gather data on the results of trials, evaluate results, and use that information as the basis for building a full consensus.

Stop, Do a Process Check: Using the Group as a Resource to Determine and Resolve Its Consensus Achievement Hang-ups

This is a technique that probes the heart of the matter by asking those involved why consensus is not being reached and what can be done to achieve it. My experience is, when given the opportunity to step back together and look at why they aren't meshing, group members have keen insight. This self-examination process should be facilitated in an open, collaborative way, with conflicts and disagreements encouraged, but managed so that they stay constructive. The key is getting the members to explore ways of overcoming any identified barriers so consensus can be achieved.

If the problematic issue lies at the core of the group's ability to function—politics, mistrust, an unyielding blocker, and so forth—you are at a crossroad. Under these conditions, achievement of consensus may be impossible. You then can decide whether you want to face the deeper problem. If not, at least you know what the barrier is. You can cut your losses and stop wasting time on a situation that doesn't have the ingredients to make a consensus possible. Or, you may be forced to move ahead without consensus by resorting to one of the techniques described next.

SIX WAYS TO MOVE AHEAD WHEN CONSENSUS IS NOT POSSIBLE: MAKING PROGRESS WITHOUT CONSENSUS

Consensus is never a certainty. In some cases a problem will be so difficult, or the hostilities so great, that even though you are an able facilitator who has done all of the right things, consensus just could not be achieved within the time frame before the deadline. In other instances, all parties involved will genuinely put everything they have into trying to reach a consensus, but just not be able to do so. In such circumstances where consensus is not possible at the time, there are several actions open to keep you from giving up and doing nothing. Since none of these actions is consensus, they are less optimal actions, but they do continue momentum and progress.

Determine a Fallback Option under Fire

This is not a good thing, but it happens. Having made the mistake of not setting a contingency option in place before starting the process, you now have to figure out the fallback option in the crunch. In other words, in the heat of battle, under severe time constraints and pressures, you now are forced to determine how the decision will be made.

If you are the manager of the group, holding a position of authority means you can choose to make a CA II executive decision, call for a vote, or send the problem to a higher authority for a decision. If you are a chairperson without position authority, quite often the only possible alternatives are the latter two. The problem with this approach is that it means playing the game of surprise, and often the chosen decision option is not readily accepted by the group.

Go to Your Predetermined Fallback Option

Bring into play the fallback, or contingency, decision option decided upon in advance of the deliberations: CA II executive decision, vote, or elevation to a higher authority. The importance of having a contingency option declared before debate and discussion begins was well documented in the previous section. While not ideal, it gets the decision made and does not destroy team spirit, because everyone knew this option was a possibility from the beginning of the consensus process.

Go with the Solution Favored by the Implementers

Suggest that since the group has not reached consensus and yet action must be taken, the group should proceed with the solution favored by the members who have the major responsibility for implementing it, since it's up to them to make it work.

Use a Nonprecedent-Setting Majority Vote

Suggest that since consensus was not possible, the group should consider moving forward with the option favored by a majority vote, but with a stipulation that this alternative does not set a precedent and cannot be used as a basis for future decisions.

Ask for a "Stand Aside"

Any refusals to support a preferred solution should be carefully examined and processed. Effort needs to be placed on truly understanding the conviction level of the objector as well as bias and self-interest. However, pressure to force the person to conform must be avoided. If an acceptable proposal cannot be found that brings the objector on board, then he should be asked if he is willing to step aside, not interfere with implementation, but comply with the outcome where necessary. If the objector is willing, then the decision is made.

An objector may elect to stand aside during a decision when personally unable to support it but sees no irreparable harm coming to the group if the decision is made; in other words, there is no need to protect the group from its own decision. But as the leader, be aware if several objectors want to stand aside. This is not a good sign. It most likely indicates that the present decision is not fully formed in some important ways and concerns are lingering. It is best then to continue discussion to address the concerns before moving ahead with a decision. Use the "stand aside" approach rarely and with caution, as it gives a special "pass" to a team member and doesn't make them part of the consensus.

Override the Dysfunctional Blocker

If consensus has been thwarted by a chronic objector who, in spite of everyone's sincere efforts to understand and accommodate his interests, still digs in with a "Hell no, I won't go" attitude—making it clear "I do not agree," "I will not support," and "I will not stand aside"—then the will of the majority should determine the decision. However, you must realize that this is a sensitive situation requiring careful facilitation.

A group manager or another person with the authority to render a final decision might say something like, "Well, in spite of our extensive discussion, consensus is not possible at this time. Since the deadline for the decision is today, I'm going to have to make the decision myself. Is there any other information that anyone wants to share with me before I decide?"

Any information that is offered would be listened to and confirmed. Then the manager would make a decision that conforms to the wishes of the majority—that is, everyone but the chronic objector.

Similarly, a task force or committee chairperson who lacks the ability to make an executive decision could say, "Well, in spite of our extensive discussion, consensus is not possible at this time. Since the deadline for the decision is today, we are going to have to vote. Is there any information that anyone wants to share before we vote?"

Any information that is offered would be listened to and confirmed. The vote would be held. The chronic objector would be outvoted and the wishes of the majority would prevail.

The point is, a whole group cannot be held hostage by one unyielding person who discounts the interests and needs of others despite all the genuine efforts to accommodate her feelings, interests, and desires as various potential actions are developed.

As a collaborative leader, the skill of "consensus molder" must be in your repertoire. After all, no decision option hinges more on genuine collaboration than that of consensus.

CONSENSUS BUILDING: KEY LEARNING POINTS AND WHAT I WANT TO DO DIFFERENTLY TO IMPROVE MY SKILLS IN THIS AREA

My Key Learning Points from Chapter Six:

What I Want to Do Differently:

READER REFLECTIONS AND APPLICATION ACTIVITIES FOR CHAPTER SIX

Reflections

Regarding my facilitation of consensus decision making, do I ensure that:

1. All teammates are getting an equal chance to participate, to have their views heard, understood, valued, and written down?
2. All teammates are working to find the best in others' ideas instead of attacking those ideas to make their own look better?
3. All teammates are involved in influencing the final decision so everyone is more likely to collaborate in implementing it well?
4. All teammates understand why this decision was made and why other decisions were not made?

Leader Application Activity: Consensus Building with an Individual (30 to 35 minutes)

1. Have teammate read Chapter Six and come prepared with answers to two questions in advance of meeting with you:

- What did you feel were the most important learning points from the chapter?
- Why were these key learning points for you?

2. Open session by letting the teammate share his or her views on the two prework questions above. Actively listen, understand.
3. Share what you found important in the chapter and why these points were key to you.
4. Have the teammate present a real work situation where they used the consensus option. Examine results in terms of the Figure 6-1 model, the six standards, or the eight principles. Could any of those have helped the process or results? Which ones? In what ways?
5. Have the teammate present a decision he or she is facing where consensus would be the best decision option. Mutually develop a plan incorporating any of the ideas from Chapter Six for converging on win-win.
6. Hold a postmeeting follow-up. Was a genuine consensus reached? Regardless, whether yes or no, what were the lessons learned?

Leader Application Activity: Consensus Building with Your Team (45 minutes)

1. Have all teammates read Chapter Six and come prepared with answers to two questions in advance of the meeting:

   - What did you feel were the most important learning points from the chapter?
   - Why were these key learning points for you?

2. Open the session by asking teammates to share their views on the two prework questions. Actively listen, understand.
3. Share what you found important in the chapter and why these points were key to you.
4. Discuss how consensus decision making could be applied to improve team sessions.

   - Do we practice consensus here the way the chapter outlines it? Where is our process similar? Where does our process fall short?
   - Give specific decision examples where we have tried to use consensus and failed. What could we have done better with ideas from the chapter? In what ways would they have helped?

- Give specific decision examples where we have tried to use consensus and succeeded. What do we want to reinforce? What will be the benefits of doing so?
- Have we underutilized consensus decision making on our team? Is it a process we should use more routinely? Why or why not?
- Using the "Shared Values" statement from Chapter Six as a base, would it be helpful for us to adopt or refine it for our use? If yes, let's set a time and place right now to do it!

5. What types of decisions, or at what major junctures along the way to a critical decision, do we feel the consensus option would be appropriate?

NOTES

1. M. Doyle and D. Straus, *How to Make Meetings Work* (New York: The Berkley Publishing Group, 1976), 57–58.
2. A. S. Grove, *High Output Management* (New York: Vintage Books, 1985), 91.
3. Doyle and Straus, *How to Make Meetings Work,* 244.

Chapter Seven

CONFLICT MANAGEMENT

Facilitating Seven Steps to Collaborative Conflict Resolution

CHAPTER OBJECTIVES

- To examine the constructive and destructive effects of conflict on a team's output
- To review five conflict management strategies
- To present striking research underscoring the power of the collaborative strategy
- To present a seven-step collaborative model for managing conflict
- To provide the necessary how-to tips and techniques along with a case example for effectively facilitating through a conflict and on to consensus

INTRODUCTION

An old sea captain and his chief engineer got into a major conflict over which one of them was more important to the ship. Neither would concede a thing to the other. Their voices grew louder, their tones more sarcastic, and their attacks more personal. Failing to agree, they resorted to a unique plan of swapping places. The chief ascended to the bridge and the captain went into the engine room.

After a couple of hours the captain suddenly appeared on deck covered with oil and soot. "Chief!" he yelled, wildly waving aloft a monkey wrench. "You'll have to come down here; I can't make 'r go!"

"Of course you can't," replied the chief. "We're aground."

Destructive conflict has struck again—big egos, oneupmanship, personal attacks, foolish decisions, and disastrous consequences. It doesn't have to be this way, as you'll see.

This chapter on collaborative conflict management is in many ways an extension of the behaviors, tools, and processes covered in the previous consensus chapter. Taken together, the two chapters present a host of simple but powerful approaches to include in your collaborative leadership arsenal for building collaborative partnerships—especially when the situation you are facing is conflict-filled and you're trying to uncover win-win outcomes.

THE ESSENCE OF CONFLICT MANAGEMENT: SOME INITIAL POINTS

Anyone in a leadership role will sooner or later have to deal with conflict. This is a normal and essential dynamic of teams and teamwork. If all members' approaches, perspectives, and values were the same, there would be little need for group decisions at all—certainly there would be virtually no need for conflict facilitation. The very idea of conflict management, in fact, presupposes divergent ideas, opinions, and proposals regarding the best course of action or method for solving a common problem, reaching a goal, or making a decision. The goal of the team should not be to avoid or eliminate conflict, but rather, to view this diversity as a healthy and essential step in the process of working out a constructive win-win resolution to the problem at hand.

Many leaders, managers, and chairpersons experience their most uncomfortable moments at the helm when attempting to facilitate interpersonal differences and conflict. Frequently such situations create immense difficulty for them because they don't have the foggiest idea how to facilitate through the disharmony and help the group build a win-win solution. The advice in this chapter will help you rectify that problem.

Conflict in Perspective

Conflict is a word used in many ways. For our purposes the sidebar definition will be helpful.

The parties can be individuals, groups, organizations, or nations. Their wants may range from acceptance of an idea to control of a limited resource. Our definition specifies that conflict is a condition that arises when seemingly incompatible desires clash. It may be a temporary condition or one of long duration—a mild skirmish over a routine matter or a complex, highly charged interaction. It may result in visible, vigorous activity or internal ferment.

> Conflict occurs when two or more parties discover that what each wants or values is incompatible with what the other wants or values. An incompatible want or value is one that interferes with or in some manner hinders the achievement of the other.

Conflict falls into two categories: (1) *substantive conflict,* which occurs when participants are in opposition relative to the content or substance of the issues under discussion; and (2) *personal conflict,* which derives from the emotional clashes that occur during the struggle to resolve the issues at hand.[1] This chapter will describe methods to collaboratively handle substantive conflict even in situations where feelings and personal issues may arise to make management of the conflict more difficult.

When differences exist, destructive conflict may or may not be present. Differences—particularly differences in values—often lead to destructive conflict. Whether differences actually do lead to destructive conflict depends on the character and facilitation of the interaction process. In essence, your challenge as a collaborative leader is to encourage diversity without encouraging personal conflict—to harness the constructive power of substantive conflict without igniting its destructive power.

Differences as a Constructive Force

There are four good reasons for encouraging "fruitful friction" among teammates as they process information during a group session.

1. *Critical thinking is stimulated.* When an individual or a contingent challenges the direction of a team or takes exception to an offered proposal, the team is forced to reexamine its own beliefs in some detail and to reconsider previously ignored or skimmed over aspects of the issue.
2. *Innovation and creativity is sparked.* When people are in conflict over acceptable alternatives, this diversity can motivate the team members to work out new and creative alternatives that can be supported by everyone.
3. *New understanding is acquired.* Contrary arguments, opinions and ideas sharpens team members' insight and increases the breadth and depth of their understanding of the subject.
4. *Healthy debate and discussion is energizing.* The excitement and energy that spring from interpersonal differences can increase the motivation and involvement of team members in tackling the task at hand.

Differences as a Destructive Force

Disagreements and differences are destructive when they paralyze the team's ability to realize its desired outcomes. There are four reasons why this cannot be tolerated.

1. *"Winners" are produced at the expense of "losers."* Win-lose is individual selfishness manifesting itself. People's energies are directed toward each other in an adversarial atmosphere of total victory versus total defeat. Execution of

"the winning decision" is a constant struggle or even impossible because of the active or passive resistance the "losers" creatively employ.

2. *Polarization is fostered.* In a destructive mode, conflict does not produce "fruitful fiction." Instead, opposing opinions cause members to defend their ideas rather than modify them. "Getting my own way" becomes more important than discovering the ramifications of and solution for the group's current dilemma.
3. *Energy is consumed unproductively.* Preparing for battle takes time and effort. In addition, alternately defending one's own position and attacking the opponent's stance in the heat of controversy drains energy from the combatants. This energy is being siphoned off in an internal "we-they" fight rather than harnessed in a cooperative undertaking of "us against the problem."
4. *A short-term orientation takes hold.* At the destructive level, members become conflict-oriented (stressing the here-and-now differences), as opposed to relationship-oriented (accentuating the long-term consequences of their differences and the methods of resolving them).

Telltale Signs of Constructive vs. Destructive Differences

You and all your teammates need to be aware of the telltale signs of the constructive versus destructive differences highlighted in Table 7-1. When differences are constructive, they should be encouraged because they are the source of alternatives to a decision. Also, dissent is needed to stimulate the imagination, to develop the creative solution, and to get away from the foolish notion that there is only one right decision.

On the other hand, when dissent and disagreement head down the destructive path, watch out! This conflict needs to be recognized and facilitated against to prevent a toxic environment from erupting, one that will shatter teammates' collaborative spirit and working relationships, leaving the team unable to achieve its desired outcomes.

Table 7-1 Comparison Chart: Constructive vs. Destructive Conflict

Signals or Cues Indicating Constructive Conflict	Signals or Cues Indicating Destructive Conflict
High team spirit and a mutual commitment to the desired outcomes remains center stage.	Members resort to personal attacks instead of focusing on the facts and issues.
The task behavior of disagreement zeros in on issues, not people.	These initial personal attacks produce a whole series of attack/defend spirals that rob the session of its productivity.
The task behaviors of testing comprehension and summarizing are used by everyone to ensure that each other's viewpoints are understood—even though they may not be supported.	Emotionally charged oneupmanship or the same negative statements presented again and again by the same people is a sure sign of unmanaged, ruinous conflict.
Participants respond to what others actually are saying, not what they think others are saying.	Members do not listen to what others are saying; but rather, react to what they think others are saying.
The discussion stays on topic and focuses on achieving the desired outcomes.	Members dig in with unyielding attachment to their own ideas and refuse to seriously explore the merits of other proposals.
Time and energy are willingly spent modifying ideas and alternatives because members know they produce better results than any one person could produce alone.	When ideas that go against the majority viewpoint are challenged or criticized, the presenter quickly withdraws under pressure.

Intervening vs. Waiting

As a collaborative leader, your job is to pay attention not only to what is being said, but to how the group members are interacting. You are constantly faced with making a judgment about whether to intervene, or wait and let the conflict discussion proceed. This is not an easy call to make. The best advice is to monitor the constructive and destructive signs presented earlier and, if in doubt about what to do, let the discussion continue a little longer. However, if resentment starts to build and people begin to entrench their positions, step forward and actively facilitate.

FIVE PRIMARY STRATEGIES FOR MANAGING CONFLICT

This section reviews five basic strategies for dealing with conflict: avoiding, accommodating, compromising, forcing, and collaborating.[2] As will be shown, each of these approaches is useful and appropriate in certain situations. After the five approaches are briefly addressed, the remainder of Chapter Seven will concentrate exclusively on the strategy of collaborating.

Avoiding: Leave Well Enough Alone

Avoiding is retreating—either physically or mentally—from an actual or potential conflict situation. Sometimes called withdrawal, this may mean leaving the solution to fate or chance. Those employing this style simply do not address the conflict and are indifferent to the other person's needs and concerns even when there is a need to take a position.

Avoiding is often labeled "agreeing to disagree." The parties involved admit they disagree, then walk away with no resolution. They may or may not try again down the road to resolve things. Avoiding makes sense when you need more time to collect information, enlist support, or augment resources. It also is a good strategy when there seems little chance to resolve the differences in a satisfactory way because the timing is wrong, you lack the necessary power and/or influence, or the issue is trivial (from your perspective).

ACCOMMODATING: MAKE CERTAIN NO FEATHERS GET RUFFLED

Accommodating is a smoothing strategy that minimizes differences by ignoring or downplaying their importance. Accommodating concentrates only on similarities and agreement. It requires covering up and glossing over the actual discord by taking a posture that everything is pleasant, cooperative, and trouble free.

Accommodating may be productive when preserving harmony and sidestepping disruption are especially important, when the issue is more important to others than to you and you want to maintain future collaboration, when you want to build social credits for a later time, or when another teammate has superior power and is using it on you.

COMPROMISING: LET'S SPLIT THE DIFFERENCE

Compromising, as noted in Chapter Six, involves searching for solutions that bring only partial satisfaction to each of the differing parties. The people involved in compromise are each concerned with evading an outcome of personally not gaining anything from the interaction. Compromising means getting something rather than risking a win-lose struggle and maybe ending up with nothing.

Compromise can make sense when two parties with equal power are strongly committed to mutually exclusive goals, or when each party can gain more from a split-the-difference agreement than the best alternative available if no agreement is reached. Compromise also can be used to achieve temporary settlements to complex issues, to arrive at expedient solutions under time pressure, or when the goals of the parties are moderately important but not worth the effort and time required for collaboration.

Forcing: It Will Be Done My Way

Forcing is a power-oriented approach that involves threat, pressure, and intimidation to achieve your objectives without concern for the needs and acceptance of others. This is vividly demonstrated by a letter from Queen Elizabeth I to Bishop Richard Cox in 1559:

> Proud Prelate: You know what you were before I made you what you are now. If you do not immediately comply with my request [to read the English litany in London churches], I will unfrock you by God![13]

Forcing often takes place as an open, competitive battle that produces a clear victor and a clear loser. For a group, this is rarely a productive strategy, especially if the winners must work with the losers in the future.

With forcing, the content of the conflict often becomes generalized from a specific issue to an atmosphere of bitter dispute as each party's original firmness and sense of confidence becomes exaggerated and escalates toward coercion and arrogance. The image that comes to mind is Charlie Chaplin's film *The Great Dictator,* where Mussolini and Hitler are in barber chairs and each is pushing his chair a little higher in order to get "one up." Pretty soon they're both up to the ceiling yelling at each other. Forcing produces that kind of escalation.

Still, forcing can be useful in emergencies, when you know you are right, where unpopular actions must be implemented, where the organization's welfare is at stake, and when you need to protect yourself against those who take advantage of nonassertive and/or noncompetitive behavior.

Collaborating: Use the Team's Synergy and Diversity of Thought to Build Unity

This is an integrative approach; people in the conflict actively seek to satisfy their own goals *and* the goals of the others. Instead of splitting the pie, they look for ways to make the pie bigger. Collaborating is useful for finding a win-win when both sets of goals are too important to be compromised, for gaining commitment by turning concerns into a consensus, or for merging diverse insights to upgrade the quality of the result.

Table 7-2 provides a concise rundown of when to use each of the five strategies.

Table 7-2 Summary of the Five Conflict Management Strategies

Situational Consideration	Avoiding	Accommodating	Compromising	Forcing	Collaborating
Issue Importance	Low	Low	Medium	High	High
Relationship Importance	Low	High	Medium	Low	High
Relative Power	Equal	Low	Equal	High	Low, Medium, or High
Time Constraints	Medium to High	Medium to High	Medium to High	Medium to High	Low to Medium

Reprinted from: T. Baldwin, et al., *Developing Management Skills: What Great Managers Know and Do*. (New York: McGraw-Hill, 2008), 305. Used with permission.

Issue Importance: How important is the disputed issue? (High = extremely important; Medium = somewhat important; Low = not very important)

Relationship Importance: How important is the relationship? (High = critical; ongoing; one-of-a-kind partnership; Medium = somewhat important; Low = one-time transaction, for which many other alternatives are available)

Relative Power: What is the relative level of power, or authority, between the disputants? (High = boss to subordinate; Equal = peers; Low = subordinate to boss)

Time Constraints: The extent time is a constraint in resolving the dispute? (High = must resolve the dispute quickly; Medium = time is an issue, but not pressing; Low = time is not a factor)

BUILDING THE CASE FOR COLLABORATIVE CONFLICT MANAGEMENT

Besides the book's theme of collaborative leadership, there are three good reasons why the rest of this chapter concentrates on the tools and processes aimed at resolving conflict using the *collaborating* approach:

1. Collaborating, as a conflict resolution strategy, actually is higher order problem solving in action. Nothing special in terms of skills or abilities is required by the manager or chairperson when utilizing the avoiding, accommodating, compromising, or forcing methods. These are rudimentary activities that discount the conflicting parties' wants, concerns, and goals. There is no need to identify issues, work through them, and develop a solution that is win-win for everyone concerned. Effectively facilitating collaboration, on the other hand, requires more knowledge and insight, as well as time and commitment, on the part of the team manager or committee chairperson than do any of the other four methods.
2. Collaborating is emphasized here because it is the way to build consensus.
3. Collaborating is the one means to managing conflict that has consistent positive benefits as shown in research studies.

RESEARCH ON CONFLICT MANAGEMENT

Let's highlight the seminal research on managing conflict, since it drives home the case for the collaborative approach.

Two Harvard professors, Jay Lorsch and Paul Lawrence,[4] studied the use of collaborating, forcing, and accommodating in six organizations across three industries: plastics, container, and food. They concluded that managers in the two highest-performing organizations used collaborating to a significantly greater degree than managers in either medium- or low-performing organizations. Similarly, managers in medium-performing companies used collaborating to a significantly greater extent than those in low-performing organizations.

Regarding accommodating and forcing, Lorsch and Lawrence noted that while heavy reliance on collaborating is essential, a backup mode that includes some forcing and a relative absence of accommodating may be useful. They concluded that the presence of moderate forcing indicates that the managers they studied were more willing to reach some decision than to avoid the problem.

Gordon Lippitt[5] presents evidence that substantiates the conclusions of Lorsch and Lawrence regarding the greater effectiveness of collaborating (win-win) over forcing (win-lose). Lippitt compared 53 case descriptions of effective conflict resolution with 53 case descriptions of ineffective resolution. Of these, 58.5 percent of the effective outcomes were achieved through collaborating, as opposed to only 24.5 percent effective resolutions via forcing and just 11.3 percent through compromising. The results regarding ineffective resolution are even more dramatic: 79.2 percent of the ineffective resolutions were due to forcing, while none of the ineffective outcomes resulted from collaborating!

Marshall Sashkin and William Morris,[6] summarizing their literature review, found that research studies on the effects of the various strategies for managing differences confirmed that forcing and avoiding are associated with the development of negative feelings between conflicting parties, and are also associated with negative outcomes, such as decreased performance. Only collaborating is clearly related to positive outcomes in a wide range of circumstances. Bargaining (compromising) and accommodating seem to have mixed results.

David and Roger Johnson,[7] in their comprehensive examination of cooperation and competition, draw similar conclusions from the behavioral science literature. They noted that over 185 studies have compared the impact of cooperative and competitive situations on achievement. The results of these studies indicated that cooperation promoted higher individual achievement and greater group productivity than did competition.

In their study of 52 conflict cases involving middle managers, Phillips and Cheston[8] compared the effectiveness of the five conflict resolution strategies covered here. They found that (1) problem solving (collaboration) was most effective in cases involving interpersonal communication and organizational issues; (2) forcing, though used twice as often as problem solving, had a 50 percent failure rate; and (3) avoidance, used as often as problem solving, had an 80 percent failure rate.

Finally, Dean Tjosvold[9] researched the literature on cooperation, competition, and independence and drew these conclusions: 500 studies, conducted by many types of social scientists and summarized in recent reviews (including meta-analysis)... consistently indicate that it is through cooperative teamwork, much more than through competition or independence, that people communicate directly, put themselves in each other's shoes, support each other emotionally, discuss different points of view constructively, solve problems successfully, achieve at high levels, and feel confident and valued as a person.

The research evidence presented here was included to dispel doubts about the power and value of the collaborative, problem-solving approach to conflict management. The findings are all consistent in suggesting that proper application of the collaboration approach offers the greatest probability for producing results of the highest quality, while simultaneously giving the most enduring satisfaction to those involved.

But before moving ahead to lay out the tools and processes for facilitating collaboration, two points need to be stressed as key reminders. First, all five strategies can be useful and are appropriate under certain conditions; examples of appropriate use of each strategy were presented during the earlier review. And second, not all conflicts are worth resolving collaboratively. Collaboration is overemployed if seeking resolution to a conflict drains time and energy more wisely spent in other ways. Keeping those points in mind, let's look at how to manage conflict collaboratively.

A MODEL FOR COLLABORATIVE CONFLICT MANAGEMENT: TWO PHASES WITH SEVEN STEPS

The process model for resolving conflict is composed of seven steps broken into two phases: *differentiation* and *integration.*

Before effectively resolving conflict, you must first facilitate the differentiation phase in order to understand the nature of the conflict. This entails four steps: (1) clarifying the existing positions, (2) clarifying interests behind the positions, (3) keeping people separate from interests and positions, and (4) highlighting areas where interests are in agreement, and then areas where interests are in disagreement.

Next, you will need to facilitate the integration phase. This requires three steps: (1) building options emphasizing mutual gain, (2) evaluating the set of created options, and (3) selecting the best options—those providing the greatest mutual gain by building on the commonly held interests while resolving the most critical differing interests.

The significance of the differentiation and integration phases in conflict resolution is explained by Richard Walton:

> The basic idea of the *differentiation phase* is that it usually takes some extended period of time for parties in conflict to describe the issues that divide them and to ventilate their feelings about [the issues] and each other. This differentiation phase requires that a person be allowed to state his or her views and receive some indication that these views are understood by the other principals.
>
> In the *integration phase* the parties appreciate their similarities, acknowledge their common goals, own up to positive aspects of their ambivalences, express warmth and respect, or engage in other positive actions to manage their conflict. . . . The potential for integration at any point in time is no greater than the adequacy of the differentiation already achieved. To the extent that the parties try to cut short the differentiation phase, dialogues are likely to abort or to result in solutions that are unstable.[10]

Figure 7-1 shows our full process model. It provides a straightforward set of seven basic guidelines, organized into two phases that can be employed time and again in a variety of situations. As you utilize the model, a few suggestions from a personal standpoint will do you well in your collaborative leadership role. Have patience: a meaningful resolution may take longer than you first thought. Never corner people or personally attack them, but do assist them in saving face. Be empathetic; put yourself in their shoes. Avoid self-righteousness. And stay positive.

Let's cover the specifics needed to facilitate the seven steps of our conflict-managing sequence.

Figure 7-1. Process Model for Collaborative Conflict Management

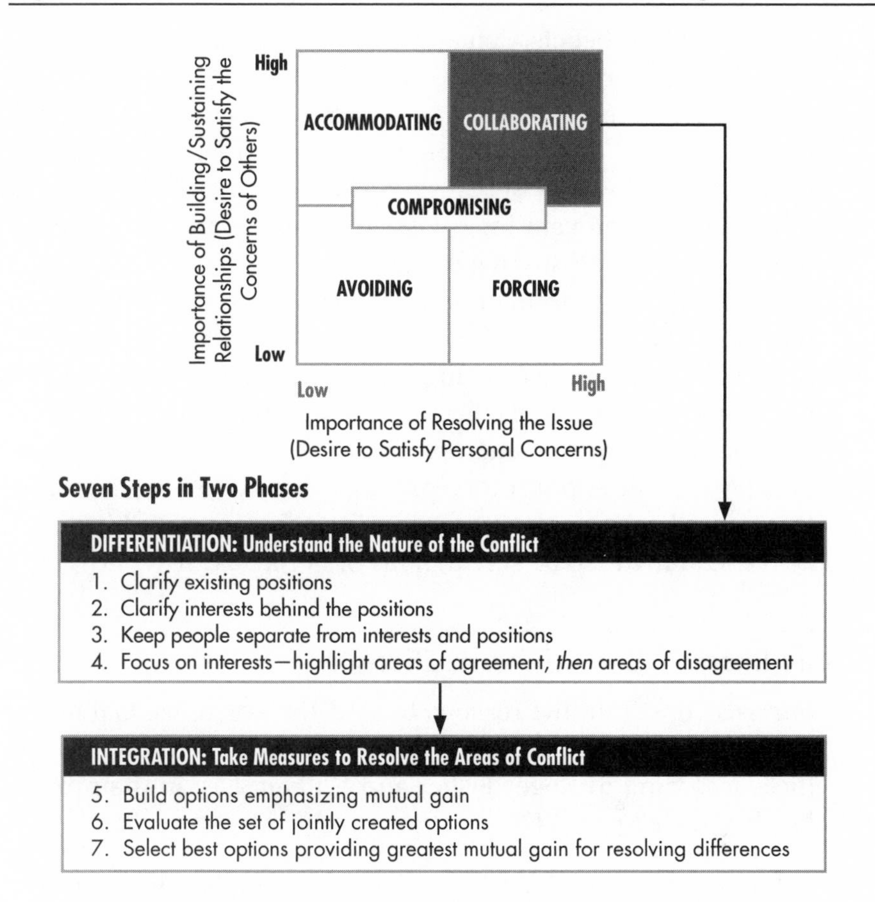

DIFFERENTIATION PHASE: UNDERSTAND THE NATURE OF THE CONFLICT

The first key to effectively managing conflict is to get a handle on where people are coming from and why. Probe with nonevaluative questions, respect teammates' rights to hold whatever views they communicate, and be sensitive to verbal and nonverbal cues. Doing so will give you the foundation for both truly understanding the *differences existing within the current situation* and the *areas of possibility for resolving the conflict.*

Clarify Existing Positions

Positions are the beliefs, demands, and assertions that each teammate holds about the issue in question.

As a collaborative leader, the first order of business is to ask your teammates to collect their thoughts and write down their positions in 20 words or less on a sheet of paper. The 20-word restriction will prevent long-winded rambling and monologues during the position-sharing process.

Next, solicit a teammate to be the scribe. Assign that person the role of recording on flip chart pages each teammate's current position on the issue being confronted exactly as it is stated, without names or initials. Each teammate isn't required to present a new or unique position statement. Any teammate may pass or acknowledge support for a previous item by asking the scribe to place a check mark by it. As an aside, the person scribing is an active teammate, so be sure that his or her ideas are noted.

Clarify Interests Behind the Positions

Interests constitute the reasons behind the positions. Gathering and understanding interests is the cornerstone for resolving conflicts according to Roger Fisher and William Ury in their excellent book, *Getting to YES:*

> Interests define the problem. The basic problem lies not in conflicting positions, but in the conflict between each side's needs, desires, concerns, and fears. . . . Such desires and needs are called interests. Interests motivate people; they are the silent movers behind the hubbub of positions. Your interests are what caused you to decide [to hold the positions you hold].[11]

Experience shows that it is easier to establish agreement on interests because they tend to be broader and multifaceted. So, digging for interests involves redefining and broadening problems to make them less rigid and hostile sounding. It allows teammates to better understand each other's point of view and to place their own views in perspective.

The process for clarifying interests is similar to the previous step. Ask the teammates to collect their thoughts and write down the interests behind the positions they hold. Again, limit this to 20 words or less. Post blank flip chart sheets next to the pages with the positions already noted. Then, using a different scribe, solicit each teammate's interests that coincide with stated positions. Have the scribe write these without any identification marks. Getting the interests recorded and posted for all to see is integral to understanding the true nature of the conflict before attempting to reconcile it.

Fisher and Ury present a delightful story in the sidebar illustrating how grasping the interests behind positions is the main enabler to resolving conflict.

The positions in this conflict were clear. One wanted the window open, the other wanted it closed. Only after the librarian asked "why" was she able to get behind the recalcitrant positions held by the two men and identify their underlying *interests.* Knowing their interests gave her the insight to come up with a solution that satisfied both men—fresh air without a draft. Clearly a win-win solution.

Consider the story of two men quarreling in a library. One wants the window open and the other wants it closed. They bicker back and forth about how much to leave it open; a crack, halfway, three-quarters of the way. No solution satisfies them both.

Enter the librarian. She asks one why he wants the window open: "To get some fresh air." She asks the other why he wants it closed: "To avoid the draft." After thinking a minute she opens wide a window in the next room, bringing in fresh air without a draft.

This story is typical of many negotiations. Since the parties' problem appears to be a conflict of positions, and since their goal is to agree on a position, they naturally tend to think about positions—and in the process often reach an impasse.[12]

"Why?" is a powerful question that uncovers the real reasons positions are taken. Probe for these deeper reasons. Discovering underlying needs and goals here will open a world of potential alternatives when the solution generation step comes up during the integration phase.

Keep People Separate from Positions and Interests

Now having clarified positions and interests, it is useful to remind everyone that this was simply an "information gathering and understanding" exercise. While the first two steps were necessary and important, from now on attention must be focused on the real issue at hand: resolving the conflict. Interpersonal confrontations are more likely to end in an outcome of mutual satisfaction if the parties depersonalize their disagreements by seeing each other as collaborative partners presenting their viewpoints, rather than as self-promoting rivals advocating dug-in positions.

Eliminating identifying marks when building the "positions and interests" flip chart pages was crucial because the objective from this point forward is to understand positions and interests as a pool of information generated by the whole group, on behalf of the whole group, to be processed by the whole group.

Obviously, some people will remember who said what relative to the flip chart material. However, visually reinforcing the recorded positions and interests with the names or initials of the information providers will only strengthen their bond to their personal viewpoints and perspectives. People often interpret criticism of their positions or interests as criticism of themselves. They react defensively, and in turn attack the positions and interests of others.

You can add to the power of the separation process by not referring to an item as "Rachel's position" or "Jorge's requirements." Instead, reference them as "one of our positions" or "one of our interests"; refer to the total flip chart information as "our team's output." Statements such as these keep subtly reminding all present that this is a teamwide collaborative effort for the good of the whole team.

Focus on Interests: Highlight Areas of Agreement, Then Areas of Disagreement

Agreement. The center of attention for this task is strictly the flip charts containing the interests information. First ask teammates to read all the posted information regarding interests, and then ask them to itemize and link those where interests are shared.

As agreements are highlighted, the scribe can circle the statements or record them on a clean flip chart page. This fosters collaboration by focusing teammates' attention on what they share in common. By making salient shared interests, everyone is sensitized to an existing base of common interests that gives momentum to resolving differing ones.

Disagreement. With the areas highlighted where interests are in agreement, the next job is to clarify the essence of the differing interests. Without bringing differences into focus, teammates are not likely to reach an outcome that truly resolves the conflict and leaves everyone satisfied.

The process is straightforward. Ask all teammates to review the pages of uncircled interests from the foregoing step and bring out the points of contention among these items. The scribe compiles a list of the divergent interests as the teammates develop it. Give recognition to teammates who demonstrate clear thinking here.

As an aid to defining and clarifying the areas of difference, the following framework can be used as a universal starting point. My experience indicates that in many instances the conflict can best be understood by assessing the dispute in terms of goals, roles, procedures, relationships, limits, timing, information, or values.

1. *Goals.* Is there conflict over what end result(s) people are trying to accomplish?
2. *Roles.* Is there conflict over who can or should do what?
3. *Procedures.* Is there conflict over the techniques, methods, systems, strategies, or tactics used for doing something (accomplishing a goal, managing a change, solving a problem, making a decision, and so on)?

4. *Relationships.* Is there conflict over how the people will relate to each other?
5. *Limits.* Is there conflict over what is or is not possible, what the group's authority or responsibility really is, what resources are available?
6. *Timing.* Is there conflict over when things should be accomplished or decided?
7. *Information.* Is there conflict over the facts, figures, or data being discussed?
8. *Values.* Is there conflict over what is right, wrong, fair, ethical, or moral?

For example, facing a conflict over creating an in-house attitude survey questionnaire versus purchasing a canned (standardized) questionnaire from a consulting firm, the HR manager could summarize the highlighted differences by saying something like this:

"All of our differences are rooted in three areas: procedures, limits, and information. Regarding procedures, there is substantial disagreement over how best to evaluate our two alternatives. As to limits, we have a dispute over whether we have the knowledge and skills in-house to create a reliable and valid attitude survey questionnaire. Finally, with respect to information, we are arguing—without any evidence one way or the other—whether it would be possible to add eight to ten questions of our choosing to a canned survey questionnaire."

By utilizing this framework, you quite often can help the group members discover that all of their points of contention can be consolidated into one, two, or at most, three of these eight categories. Helping the group members see the bare bones of the conflict makes it easier for them to resolve it collaboratively. However, since no framework can cover all contingencies, the group might have to create other categories to account for disagreements and differences that fall outside the eight presented here. Still, there is no better place to start than with this framework.

INTEGRATION PHASE: TAKE MEASURES TO RESOLVE THE AREAS OF CONFLICT

Three integrative activities aimed at maximizing mutual gain is the goal of the second phase of our process model to resolve conflict collaboratively. This phase is where you will be at the forefront in your role as collaborative leader building a united front among teammates to find a consensus solution to the conflict problem.

Build Options Emphasizing Mutual Gain

This step is the "thinking up" activity. It should be fun and stimulating for the entire group. Focus the teammates on the task at hand by reminding them that they all need to concentrate on coming up with as many creative, original, and unusual solutions that will reconcile part or all of the differing interests noted on the flip chart in the previous step.

By focusing all teammates' attention on brainstorming solutions that reduce or close the gaps among the divergent interests, the interpersonal dynamics naturally shift from competitive to collaborative. In addition, the more options and combinations there are to explore, the greater the probability of finding common ground.

By brainstorming, all teammates offer every single idea they can without worrying whether ideas are good or bad, whether they will work or not. At this point ideas should not be evaluated. The purpose here is for you to create a criticism-free atmosphere that encourages everyone to be creative and to express spontaneously a variety of options emphasizing mutual gain without fear of reprisal.

Appoint a new scribe, and as ideas are brought up, have the scribe note them on the flip chart; assuming the scribe is a teammate, be sure not to overlook that person in the brainstorming process. Once the joint search for options is concluded, clarifying questions about particular items are handled, and the list is refined and consolidated to eliminate duplications and overlaps among any of the proposed solutions. This new list is rewritten on clean flip chart pages as the final options for evaluation.

Evaluate the Set of Jointly Created Options

This step reinforces the shift in thinking from "getting what I want" to "deciding what makes most sense," and continues to foster a trusting attitude among teammates. It encourages whole business thinking while breaking the chains of overconfidence or overcommitment to one's initial position.

MaxList (Maximum List) voting is an excellent technique for taking the team's final list of options and separating the significant few ideas (the top three to five alternatives) from the trivial many. By using MaxList voting, the team concentrates on finding good and acceptable solutions, rather than on eliminating those thought to be bad and unacceptable.

With MaxList voting, each participant is given a set number of votes equal to 1.5 times the number of items to distribute across all the items on the newly scrubbed list. For example, say you have 16 items on your final list, then 1.5 x 16 = 24 votes allotted to each person. If there is an odd number of items, round upward to the next even number and then multiply by 1.5 to determine the number of votes per person.

There are three basic ground rules for MaxList voting: (1) no more than one-third of a person's votes can be applied to any one item, (2) all votes must be allocated, and (3) no fractional votes.

Staying centered on the significant few solutions considered most acceptable is a positive approach and is less likely to elicit defensive behavior. Also, this process eliminates wasting time processing obviously inadequate solutions not supported by any group member, and at the same time it avoids the divisive effect of majority rule.

Select the Best Options Providing Greatest Mutual Gain for Resolving Differences

Now, in the role of collaborative leader, focus the group on joint evaluation of the significant few options the MaxList voting process just uncovered. If a solution (or combination of them) doesn't immediately jump out, there are four good tools you can take advantage of to arrive at a consensus resolution.

1. Examine the significant few options from the "best fit" perspective by asking: "How well does each alternative overcome the incompatible interests identified by us in the second step?"
2. Zero-in on the "likes and concerns" of the significant few potential solutions by saying: "Let's build a 'likes and concerns' list for each of our top few ideas using two ground rules: (1) before you can voice a concern about an item, you must first say what you like about it, and (2) when you do raise a concern you must give a suggestion for overcoming the concern."
3. Focus on examining the mutual gains inherent in each of the significant few by asking teammates: "How does each alternative constitute a meaningful improvement for us over what exists now?"
4. Create some new alternatives from the significant few by asking: "How might we combine and reformulate the significant few solution ideas we have into new and better mutual gain solutions?"

These approaches assure a balanced examination of each of the top few proposals for reconciling the existing differences. They give you a means to help teammates guard against the normal human tendency to think only of objections and restrictions without trying to figure out how they can be overcome. They also eliminate the tendency of people to take a black-and-white position either "for" or "against" each option without realistically trying to weigh alternatives.

In essence this evaluation process forces group members to "turn their individual kaleidoscopes." When they have to look for the good in all of the stated options, group members are forced to break away from fixed positions and proprietary interests and, instead, to become collaborative partners considering several different viewpoints in conflict resolution.

Consensus will be the rule rather than the exception when the seven-step model for resolving conflict is utilized. The team is facilitated through a planned sequence of steps designed to generate authentic collaboration each step of the way. You and the team become a cohesive collaborative resource for resolving disputes, rather than you being a dictator pushing teammates to accept your solution or retreating to some half-baked compromise that leaves everyone less than satisfied.

While consensus is often achieved using the collaborative model, it is not guaranteed. Fortunately, acting as collaborative leader, you can employ a number of other consensus building techniques to get a resolution by drawing upon Chapter Six. Consensus is highly relevant to the task of facilitating the selection of a win-win option for resolving conflict.

PRISON GUARDS' UNIFORMS: A CASE APPLICATION OF COLLABORATIVE CONFLICT MANAGEMENT

Alan Filley provides an account of a group session held to decide on prison reforms in Wisconsin. His case description provides a practical application of our collaborative model for managing conflict.

Let's analyze this case by contrasting the two processes used to handle the differences over uniforms. In the first situation, the group engaged in clear-cut conflict, which was only partially resolved by vote. In the second situation, the consultant facilitated the group in a true resolution of the conflict by following the process model advocated here.

Differentiation: Understand the Nature of the Conflict

Clarify existing positions. Six guards held the position of maintaining the practice of guards wearing uniforms. Three guards held the position of doing away with uniforms.

Nine of the state's top prison officials met to design an ideal correctional institution. In the course of the discussion, one group member proposed that uniforms traditionally worn by prison guards be eliminated. The group then began a lengthy argument about whether uniforms should be worn. One group member suggested that the issue be resolved democratically by vote. As a result, six people voted against uniforms and three voted in favor of them. The winning members looked pleased, while the losing members either got angry or withdrew from further discussion.

A group consultant present at the time suggested that the members take another look at the situation. Then he asked those in favor of uniforms to clarify why they favored them. They stated that part of the rehabilitative process in correctional institutions is teaching people to deal constructively with authority, and they saw uniforms as a means for achieving that goal. When asked to clarify why they opposed uniforms, the other group members said that uniforms created such a stigma, guards had additional difficulty laying to rest the stereotypes held by inmates before they could deal with them on a one-to-one basis.

The consultant then asked the whole group what ways might be appropriate to meet the combined goals, namely, teaching people to deal with authority and avoiding the difficulty of stereotypes held about traditional uniforms. While working on the problem, the group generated 10 possible solutions, including identification of prison personnel by name tags, by color-coded casual dress, or by uniforms for guard supervisors but not for guards in constant contact with prisoners. After discussing the various alternatives, the group decided upon the third solution.[13]

Clarify interests behind positions. The guards with the position favoring the retention of uniforms said they were interested in teaching the prisoners to deal constructively with authority; those who wanted to abolish uniforms said they were interested in eliminating uniforms because they created such a stigma that guards had additional difficulty laying to rest the stereotypes held by inmates before they could deal with them on a one-to-one basis.

Keep people separate from interests and positions. There is no case information on this point.

Focus on interests: highlight areas of agreement, *then* areas of disagreement. All the guards agreed they were interested in designing an ideal correctional institution. The consultant then helped them realize that their interests were in conflict over *goals*; that is, there was conflict over what end result(s) the two coalitions were interested in achieving. Coalition one had a goal of wanting to teach the inmates to deal constructively with authority; coalition two had a goal of wanting to eliminate the negative stereotypes and difficult interpersonal relationships traditional uniforms fostered.

Integration: Take Measures to Resolve the Areas of Conflict

Build options emphasizing mutual gains. The consultant then shifted the focus from where the guards were opposing each other to where they were united in opposing the issue and mutually resolving the difference. This was accomplished by asking the nine guards to collaborate in searching for ways to meet the combined interests of both coalitions so both goals could be achieved. The nine guards collaboratively generated 10 potential solutions.

Evaluate the set of jointly created options. The 10 options were then evaluated to come up with the significant few. Three options rose to the top: identification of prison personnel by name tags, or wearing color-coded casual dress, or uniforms for guard supervisors but not for guards in constant contact with prisoners.

Select best option providing greatest mutual gain for resolving differences. After discussing the various alternatives, the guards reached a consensus to have guard supervisors wear uniforms, but not the guards who were in constant contact with prisoners.

The interests of both coalitions were met, and both sets of goals were achieved. It was a win-win solution resolving the conflict.

SUMMARY

The following remarks from the *Handbook of Conflict Resolution* provide a fitting close to our chapter on collaborative conflict resolution:

> There is a skeptic's critique often heard out in the field or in training that says, "This is fine on paper, but it isn't realistic." In detail it goes something like this. People who are angry don't want to solve problems, work with each other, or talk to each other. What may help to persuade those [skeptics] holding this view of conflict problem solving is that research and practical experience are firm in their conclusions. People *do* respond better to problem solving than the settlement-oriented ones [like forcing, compromise, or voting], they reach agreement more often and faster, they report being more satisfied, and both agreements and satisfaction hold up better in the long run.[14]

And our seven-step collaborative conflict resolution model is an excellent process for guiding you and your teammates in doing just that.

CONFLICT MANAGEMENT: KEY LEARNING POINTS AND WHAT I WANT TO DO DIFFERENTLY TO IMPROVE

My Key Learning Points from Chapter Seven:

What I Want to Do Differently:

READER REFLECTIONS AND APPLICATION ACTIVITIES FOR CHAPTER SEVEN

Reflections for the Leader

As a collaborative leader, practicing excellent listening skills is critical in conflict management. When using the seven step model presented in this chapter, demonstrating "listening for understanding" about teammates' needs, desires, feelings, values, concerns, and fears is fundamental to your facilitation process. Without excellent listening on your part, you will miss the true intentions behind your teammates' statements.

Reflect on these six questions to better understand how good of an active listener you are. Make a strong mental note to improve in any area where you need to do better.

1. Do I give the person that is speaking my total attention?
2. Do I make certain I focus on their words instead of thinking about what I am going to say next?
3. Do I maintain eye contact with the speaker at all times, even when I disagree?
4. Do I physically react to the conversation using nonverbal cues—nodding my head, smiling, waving to keep the person talking to show that I am listening?
5. Do I use verbal encouragement phrases like "tell me more" or "help me to understand what you mean" to probe and learn more about what is on a teammate's mind?
6. Do I use "I" messages not only to help to clarify the thoughts and feelings of others, but also as a way to convey my thoughts, opinions, and needs in a nonthreatening way? Using "I" messages simply involves expressing your words, thoughts, needs, and feelings in a straightforward statement that begins with "I," like: I am concerned about," "I would prefer if we would," "I suggest we think about," "I am hearing you say," or "I am sensing that you."

Leader Application Activity: Conflict Management with an Individual (30 to 45 minutes)

1. Have teammate read Chapter Seven and come prepared with answers to two questions in advance of meeting with you:

- What did you feel were the most important learning points from the chapter?
- Why were these key learning points for you?

2. Open the session by letting the teammate share his or her views on the two prework questions above. Actively listen, understand.
3. Share what you found important in the chapter and why these points were key for you.
4. Have the teammate present a past work situation where he or she was engaged in a conflict situation. Examine results in terms of each of the seven steps of the conflict model. Could employing the model have improved the process or results? Which steps would have made a difference? What might have happened differently?
5. Have teammate present a current situation where conflict may be brewing. Mutually review the eight-dimension framework (goals, roles, relationships, etc.) to understand the conflict source(s). Then mutually develop a conflict resolution strategy using the Figure 7-1 conflict model.
6. Hold a postmeeting follow-up. Was the resolution experience successful? Regardless, if yes or no, what lessons were learned?

Leader Application Activity: Managing Conflict with Your Team (45 minutes)

1. Have all teammates read Chapter Seven and come prepared with answers to two questions in advance of the meeting:

   - What did you feel were the most important learning points from the chapter?
   - Why were these key learning points for you?

2. Open the session by asking teammates to share their views on the two prework questions. Actively listen, understand.
3. Share what you found important in the chapter and why these points were key to you.
4. Discuss how managing conflict collaboratively could improve relationships and/or productivity within your team (or between your team and another team you regularly work with):

- What issues or problems usually create conflict within our work team (or between our team and another)? What might be the source(s)?
- Discuss how we might resolve an existing intra- or interteam conflict through the application of the Figure 7-1 conflict resolution model.
- What interpersonal behaviors tend to escalate our conflicts? Why is that? What reactions do these behaviors bring out in opponents?
- What interpersonal behaviors tend to resolve our conflicts? Why is that? What reactions do these behaviors bring out in opponents?

NOTES

1. For a detailed treatment on how to handle emotional clashes, see: T. A. Kayser, *Mining Group Gold: How to Cash-In on the Collaborative Brain Power of a Team for Innovation and Results,* 3rd ed. (New York: McGraw-Hill Companies, 2011), Chapter 10, "Facilitating Feelings: Keeping the Gold Mine Productive in the Face of Emotion."
2. R. R. Blake, H. A. Shepard, and J. S. Mouton, *Managing Intergroup Conflict in Industry* (Houston: Gulf Publishing Company, 1964); P. R. Lawrence and J. W. Lorsch, *Organization and Environment* (Homewood, Illinois: Irwin, 1969); and, K. W. Thomas, "Conflict and Conflict Management," in M. D. Dunnette, ed., *Handbook of Industrial and Organizational Psychology* (Chicago: Rand McNally, 1976), among others, early on identified these five basic approaches to managing conflict. While terminology may differ among various authors, the behavioral aspects are remarkably similar.
3. L. D. Eigen and J. P. Siegel, *The Manager's Book of Quotations* (New York: Amacom, 1989), 76.
4. Lawrence and Lorsch, *Organization and Environment,* 78–83.
5. G. L. Lippitt, *Organization Renewal: A Holistic Approach to Organization Development* (Englewood Cliffs, New Jersey: Prentice-Hall, 1982), 152.
6. M. Sashkin and W. C. Morris, *Organizational Behavior: Concepts and Experiences* (Reston, Virginia: Reston Publishing Company, 1984), 326–327.

7. D. W. Johnson and R. T. Johnson, *Cooperation and Conflict: Theory and Research* (Edina, Minnesota: Interaction Book Company, 1989), 172.
8. E. Phillips and R. Cheston, "Conflict Resolution: What Works?" *California Management Review*, 21 (4), 1979, 76–83.
9. D. Tjosvold, *Teamwork for Customers: Building Organizations That Take Pride in Serving* (San Francisco: Jossey-Bass, 1993), 7.
10. R. E. Walton, *Managing Conflict: Interpersonal Dialogue and Third-Party Roles*, 2nd ed. (Reading, Massachusetts: Addison-Wesley Publishing, 1987), 92.
11. R. Fisher and W. Ury, *Getting to YES* (New York: Penguin Books, 1983), 42.
12. Ibid., 41.
13. A. C. Filley, *Interpersonal Conflict Resolution* (Glenview, Illinois: Scott, Foresman and Company, 1975), 33.
14. M. Deutsch, P. T. Coleman, and E. C. Marcus, eds., *The Handbook of Conflict Resolution: Theory and Practice*, 2nd ed. (Hoboken, New Jersey: John Wiley & Sons, Inc., 2006), 207.

CHAPTER EIGHT

DELEGATION EFFECTIVENESS

Increasing the Capacity of Others to Act

CHAPTER OBJECTIVES

- To distinguish the critical—but different—roles of authority, responsibility, and accountability in producing effective delegation
- To present the ABCDE process model for excellent delegation
- To review and contrast the five degrees of delegation
- To cover what the leader typically can and cannot delegate
- To share the benefits and barriers of delegation from both the leader's and the teammate's perspective

INTRODUCTION

One of my favorite stories as a little boy was "The Little Red Hen." It's the story about a hen who found some grains of wheat in the barnyard and asked, "Who will help me plant these grains of wheat?" The dog, the goat, and the turkey all said they would not help. So the Little Red Hen said, "I'll do it myself, then."

The wheat grew tall and ripe. Then, each time she asked for help cutting the wheat, threshing it, grinding it into flour, gathering ingredients, preparing dough, and baking a loaf of bread, the three other animals always declined to help. In each case her answer was always the same, "I'll do it myself, then," and she did.

So after baking the bread, its delicious smell filled the barnyard. The Little Red Hen saw the dog, goat, and turkey hanging around the bread as it cooled. She then asked, "Who will help me eat the bread?" All three animals enthusiastically agreed to help. "Oh no you won't," she said, "I'll eat it all myself," and she did.

The Little Red Hen certainly was a motivated and dedicated worker. However, she wasn't a good leader; she didn't know how to delegate and rally support for a job.

When others are reluctant to take on tasks or help out, and you decide, "Then, I'll do it myself," you are setting a very bad precedent. To be an effective collaborative leader, delegation has to be one of your eight essential skills. As a leader it is your job to get work done through and with others. The hazards of doing everything yourself will stretch you too thin and eventually lead to a serious case of burnout. On the other hand, delegating multiplies your effectiveness because you are developing the capabilities of your people, which increases the opportunities for more and better collaborative partnerships. In the long run, delegating even may help you make more "bread"!

WORKPLACE REALITIES

It is no secret that one of the best ways for teammates to develop their knowledge, skills, and abilities is to work on challenging assignments that push them to go beyond their current level of functioning. Therefore, well-executed delegation that *increases teammates' capacity to act* provides an excellent opportunity for their growth and development. Nevertheless, as Robert Quinn candidly points out, the realities of the workplace are such that delegation often is resisted and its great benefits are missed.

> [Some] managers resist delegation using arguments such as "I tried that once and the employee fouled things up

> royally" or "Delegate my authority? Why? I'm the manager—that's my job—I can do it better myself." Managers who learn to delegate effectively, however, find it results in a variety of important benefits for themselves and the organization, as well as for the employees. Of course many managers realize that by delegating some of their work, they provide themselves with additional time and thus are able to focus attention on more significant issues. . . . Just as important, managers give their employees opportunities to develop new skills and abilities, as well as to learn more about the work unit and how it functions. This not only helps employees to be more effective in their work but also strengthens the work unit, thus allowing for a better allocation of unit resources.[1]

Excellent delegation is a systematic process. The process fundamentals are clear cut, and when put into practice the results can be compelling. Lester Urwick, one of the grand masters of early management thought, understood the power of delegation over 65 years ago when he emphatically claimed that both the "lack of courage to delegate properly, and the knowledge of how to do it, is one of the most general causes of failure in organizations."[2]

AUTHORITY, RESPONSIBILITY, ACCOUNTABILITY: CORE ELEMENTS OF DELEGATION

Every time you delegate work to a teammate, three inescapable core elements of delegation are in play. *Authority, responsibility,* and *accountability* form an integrated process and must be applied by you as a unified whole. Failure to do so will wreck the delegation process and produce less than desired results.

By understanding the critical nuances of authority, responsibility, accountability, and their underlying principles, you can avoid a lot of trouble when applying delegation within your team. The three inescapable core attributes of delegation are like a three-legged stool; each depends on the others to support the whole, and no two can stand alone. Figure 8-1 shows a graphic illustration of this three-way relationship.

Figure 8-1. Authority, Responsibility, and Accountability Relationships

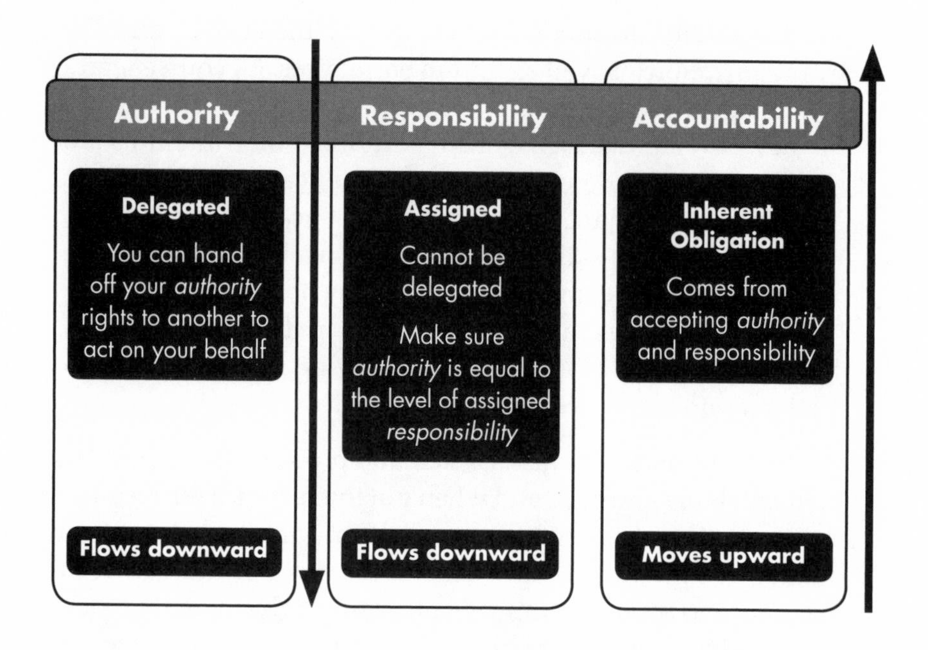

Authority *Can* Be Delegated

As a leader, you can transfer pieces of your formal authority (usually defined in a job description or a decision rights document) to another teammate when assigning a task to that person. In essence, you can deputize your teammate to take action on your behalf within the boundaries of the delegated (transferred) authority.

Authority chiefly comes from the *power of position;* it is based on your slot in the hierarchy and the defined decision-making rights that correspond to that position. Your authority determines the extent to which you can control resources, use judgment, make decisions, and take action in pursuit of your assigned objectives. The more authority you have, the greater your ability to delegate higher-level, more meaningful and challenging tasks to others to help them learn, develop, and grow.

Responsibility Cannot Be Delegated, but It Can Be Assigned

As a leader, you can assign responsibility to another teammate in terms of the results that need to be achieved. However, you need to keep in mind that you only *assigned* responsibility to your teammate. In other words, ***you created a joint responsibility*** with your teammate because you are still ultimately responsible for the execution of the assignment, project, or task.

If your teammate "fouls up the thing royally," your manager will censure you, not your teammate. In short, you can never fully hand off any of your responsibilities to someone else. Assigned responsibility should be made in terms of the goal or results to be accomplished, not the detailed specifics for doing the job. Expressing assigned responsibility this way adds vitality to the process.

Accountability Means Obligation

Accountability is the moral compulsion felt by a teammate to meet the goals and objectives of an assigned task. As a result of accepting a task assignment, your teammate in effect gives you a promise—either expressed or implied—to do her best in carrying out the activities associated with it. Having taken on the task, your teammate is obligated to complete it, and thus is held accountable by you for the results produced.

Let's use an example to highlight the relationship among these three delegation elements. Assume you—a manager—assign responsibility for achieving a particular budget target to one of your people, Debbie, and also delegate the appropriate formal authority so she truly can take charge and control her expenses. By doing so, you have created a joint responsibility with Debbie and can hold her accountable for her results. Thus, if Debbie, through her own budget mismanagement (or your mismanagement of her) ends the year 25 percent overspent, which in turn causes you to come in over budget, your manager is still going to hold you responsible and accountable for the overspending in your budget. Your manager will "ding" you, not Debbie. If Debbie finishes the year on target or better than planned, you both can take pride in fulfilling a joint responsibility.

FOUR PRINCIPLES FOR INTEGRATING AUTHORITY, RESPONSIBILITY, AND ACCOUNTABILITY

Following are four principles that will serve you well in delegating effectively and increasing the capacity to act of a person or even a small team of people.

Principle 1: True delegating involves two actions: (1) Giving up what you would most like to hold on to—the authority that carries with it the ability to command and control, and (2) Holding on to what you would most like to give up—the responsibility that carries with it risk and answerability for results. This is the exact opposite of what a manager once jokingly asked me: "Can I make all the decisions yet have his butt on the line for all the results?"

Principle 2: The amount of authority you delegate to a teammate must be commensurate with the depth and breadth of the assigned responsibility in order to give the teammate a fair opportunity to accomplish the intended results for which he will be held accountable.

Principle 3: The joint responsibility between you and your teammate for performance results is mutually unconditional. If the teammate fails to perform a certain task or does so poorly, then you share in that teammate's failure. But by the same token, if the teammate succeeds, then you share in that success as well, and your operation benefits accordingly.

Principle 4: Accountability always is a teammate's inescapable, upward obligation to you arising directly from your assigned responsibility and delegated authority to that teammate.

A PROCESS MODEL FOR INCREASING CAPACITY TO ACT THROUGH EXCELLENT DELEGATION

The process model is outlined in Figure 8-2.

Figure 8-2. Increasing Others' Personal Capacity to Act

Choose a Capable Person

Choose a *person or team* that will have a reasonable chance of succeeding in the assignment. This is the initial step before you begin the actual delegation process.

(Increasing Capacity to Act should always be success oriented)

Alignment (A)

Explain the task's purpose, provide a vision of how it fits into the bigger picture, and establish a set of shared expectations about the desired outcomes to ensure alignment.

(Alignment builds commitment and locks down a sharp focus)

Boundary Conditions (BC)

Present any non-negotiable boundary conditions that may exist relative to the task AND mutually negotiate any open boundary conditions.

(Boundary Conditions define the range of freedom)

Delegation (D)

Provide the authority and means to do the job. Delegate the necessary *authority* corresponding to the *assigned responsibility* and help the person or team acquire the resources needed to successfully meet the task's objectives.

(Delegation creates a *jointly* accountable collaborative partnership)

Examination (E)

Examine progress on an agreed-to schedule to stay in touch; and, help out, when necessary, via coaching and counseling.

(Examination ensures the teammate is not abandoned or just dumped on)

If you haven't already noticed in Figure 8-2, the four process steps for increasing the capacity to act of another person or small team is as easy as ABCDE to remember and apply:

$$A + BC + D + E = TM_{\text{Increased Capacity to Act}}$$

Alignment + **B**oundary **C**onditions + **D**elegation + **E**xamination = Teammate $_{\text{ICtA}}$

But before holding your initial meeting to share and process information around alignment, boundary conditions, and delegation elements inherent in the task to be passed on to another, you need to give serious thought to the teammate or set of teammates most capable of being successful in achieving the desired results.

Let's first examine the primary considerations in choosing the person, and then turn our attention to each of the four process steps to give you a deeper understanding of the model.

Choose a Capable Person

Whether you are delegating to get a job done or to aid in the development of a teammate, it is important that you take time to evaluate both the task and the teammate's capabilities. Candidates for delegation should be selected based on two criteria: (1) they should be qualified to receive the assignment, and (2) they should be willing to accept the delegation. Tasks should be assigned to an appropriate teammate or small group that understands, agrees with, and commits to the delegated assignment. The delegation should not be imposed. So make sure you can explain to the teammate what needs to be accomplished, along with why this assignment is important to you and your operation.

Some tasks can be done better by a teammate because that person has more expertise, is closer to the problem and can obtain more timely information about it, or because you simply do not have the time necessary to perform the task appropriately. In cases where you are using delegation to develop a person, you need to be sure the assignment is at a proper level of difficulty, providing your teammate some challenge but not so much that he becomes frustrated with the job.

Bottom line, you need to know the task requirements well and have a reasonable understanding of what the teammate is able and willing to do. There is a real need to avoid delegating too much, too soon, to someone who is not capable of handling the responsibilities at this time. As the model pointed out, increasing the capacity to act of another person or small team must first and foremost be success oriented. Success in this endeavor is one of the major builders of mutual trust.

Alignment (A)

Alignment is the first point of discussion in your delegation meeting with the individual(s) involved. It means sitting down with the chosen teammate(s) and doing three things:

1. Explaining the required deliverables of the task—to make clear what is to be achieved.
2. Reviewing how the particular task fits in with the goals of the larger unit—to give meaning to the task.
3. Covering your rationale for choosing that particular person or small team for the assignment to be delegated—to instill confidence and motivation.

As you carry out the alignment activity, remember it is a two-way discussion. If you're doing all the talking, you won't be sure the teammate or the small group buys into the idea of delegation or understands the specifics of the assignment. Ask questions, probe, and solicit thoughts and ideas from all involved. Stay open and flexible. Besides ensuring a full understanding of the task and how it aligns with the unit's higher level objectives, you want to be certain there is commitment from the participants.

While the details of your communications will vary, here is a set of alignment topics to keep your meeting on the right track. Remember, each topic is a dialogue!

- Describe the task as you see it. (Background information and basic requirements.)
- Discuss the expected results. (What defines success.)
- Talk about how the task fits in with the teammate's other objectives and how it aligns with the goals and strategies of the larger unit.
- Go over any elements that will require your hands-on coaching along the way.
- Confer on ideas to minimize conflicting priorities or other risk factors and/or concerns.
- Engage in dialogue about why you need help.
- Explain why you have chosen this particular person or small team.
- Talk about the importance and relevance of the assignment.
- Discuss the benefits for the individual or team.
- Seek full commitment before moving on.

It is important to obtain verbal commitment to the new responsibility in the alignment stage, because if it does not exist here, you need to think about another choice. The person's emotional contract with you and commitment to the project is critical to a successful outcome. And you must find this out sooner rather than later.

Boundary Conditions (BC)

Once the alignment conversation is finished, the next critical phase is to hold a thorough review and discussion of the boundary conditions. Increasing another's capacity to act in no way gives unrestricted freedom for her to operate in whatever manner she sees fit. There always will be limits involved in any delegated assignment—some non-negotiable, other's negotiable—that delineate the range of operational freedom. It is imperative that all known boundary conditions are brought up and mutually examined.

All *non-negotiable* boundaries need to be presented first and explained. These come with the assignment, and while they have to be gone over to ensure understanding, they are not open to negotiation and change. Depending on any given assignment, things like deadlines, schedule, budget, people, resources (in-house versus outside vendors), facilities, office space, equipment, and other similar factors could be nonnegotiable constraints that predefine some of the limits built into the assignment.

Negotiable boundaries are those that are open to discussion and to give and take. The final configuration for a negotiable boundary is whatever you and your teammate(s) agree to. Things like the number of checkpoint reviews to be held, who is involved in the reviews, or the number of status reports and their format are all good examples of negotiable boundaries. However, keep in mind that what is a nonnegotiable boundary in one case may be negotiable in another, and vice versa. Initiating and then working through the boundary conditions that are open to negotiation occurs only after the nonnegotiable boundaries have been understood and accepted.

As mentioned earlier, boundaries are critical because they determine the delegation playing field; that is, the range of freedom for the capacity to act. It curbs the all too familiar situation where a person or small group assumes it has more latitude than it actually does—and becomes frustrated when it is stopped from taking certain actions, or finds its final recommendations or implementation efforts prohibited from taking effect.

Relative to negotiable boundaries, there are five useful guidelines to keep in mind as you collaborate in reaching a consensus on them with the other person or a small group:

1. *Boundaries must be neither too broad nor too narrow.* If too broad, there is no sense of guidance, no sense of direction. If too narrow, boundaries become directive and restrictive; they usurp empowerment rather than guide it.
2. *Boundaries must be unique to the situation.* What was a negotiable boundary in one instance may be nonnegotiable in another. What was a boundary for one task may be irrelevant in another.

3. *Boundaries must be consistent with the knowledge, skills, abilities, and confidence of the individual or team.* Too many boundaries demotivate an experienced, ambitious person or team. Too few boundaries scare and overwhelm an inexperienced or nonconfident person or team.
4. *Boundaries must be realistic.* If "pie in the sky," the boundaries will create false expectations about what can be accomplished or how it can be accomplished. If too confining, there is no motivation for people to give their best effort.
5. *Boundaries must be renegotiated when anyone who was party to the original negotiation feels a squeeze of unease.* Negotiable boundaries cannot be changed without renegotiations between you and the others. Once renegotiated, the new or changed boundaries must be openly and honestly communicated to everyone impacted by the changes. To do less creates feelings of betrayal and, as noted in Chapter Four, it erodes trust.

Delegation (D)

This phase concerns providing the means and authority so the other person or the small team can produce the necessary results. If you are unwilling to this, or you like to delegate authority but then reclaim it at the first sign of trouble, don't bother trying to delegate. Nothing positive comes from inauthentic delegation.

First of all, lacking the proper degree of decision authority, the teammate receiving the delegation will not have a fair chance to be successful in delivering the "asked for" results. Then, except for immense frustration and less than satisfactory performance on her part, little else will occur. Now it will take nearly as much of your time, micromanaging her through pushing, prodding, and cajoling, to make up for her demotivated performance as the effort moves along. And if you do provide the right amount of authority to accomplish the assigned objectives but are quick to take it back when your doubts—quite often unfounded—cause you to hover, meddle, overturn decisions, or second-guess plans and actions, you will do nothing but create mistrust and a lack of willingness on any teammate's part to accept further delegation.

Your goal in increasing another's personal capacity to act is to maximize their influence over the task. This means delegating the most authority within the widest possible boundary conditions appropriate to the individual's competence and the demands of the task. The quality of your delegation is based on how much real influence and final decision-making power you pass on to the person to optimize their contributions in delivering the agreed-upon objectives.

You may feel you can do the job faster without bringing in a teammate and delegating to that individual. In the short run you may be right. In the long run, however, excellent delegation will enable your teammates to develop greater competencies, and will allow you more freedom to devote time to longer range matters. This will become apparent only if you have assigned responsibility and delegated authority clearly enough for your teammate to know what is expected each and every time.

Examination (E)

Upon working through the delegation piece with the teammate or small group, the final segment is to determine your role for keeping in contact. This is a negotiable condition and should be an open, honest conversation that covers both the formal methods (like checkpoints and their requirements) and the informal (like interpersonal relationships).

The actual amount of your formal contact will vary depending on your teammate's confidence, skill level, and previous experiences in handling this or similar tasks in the past. Still, above all else the central issue in delegation is *trust*. And no teammate—even someone new to the situation—will feel trusted if you suffocate him with too much supervision.

As the task unfolds, you need to keep asking yourself this question: "Within the framework of my teammate's capabilities for this assignment, as mutually probed and discussed by us, am I now really letting that person take charge; or am I restricting this individual's personal capacity to take suitable action by keeping strings attached and withholding authority I should be delegating?"

Examination is a helping activity so teammates can learn, develop, and become greater assets to you and the organization. Remember to make all your scheduled reviews collaborative in nature by maintaining your role as a sounding board, consultant, teacher, and helper as you check progress.

DEGREES OF DELEGATION

One nice thing about increasing capacity to act is that it does not have to be all or nothing. There actually are five degrees of delegation, as shown in Figure 8-3. This gives you a great deal of flexibility in delegating authority, assigning responsibility, and forming joint accountability obligations.

For example, an inexperienced teammate just joining your team might be started at delegation level one or two on a few key assignments. You would increase the degree of delegation as your teammate gains more confidence and greater experience. Also, keep in mind that a teammate can have different levels of confidence and experience for different tasks or assignments. Therefore, you may need to *use different degrees of delegation with the same person,* depending on the inherent complexities of the various tasks and your teammate's capabilities in handling each of them.

In another instance, due to pressures on you and the amount of tasks you have on your own plate, you may select a high-performing teammate who has successfully handled the type of assignment that you need to delegate now and use delegation level five with that person. You turn the task over to him to perform on his own with minimal discussion and oversight by you. Increasing capacity to act always is a mutual activity between you and your teammate or small group; and you move up and down the delegation ladder as required using the ABCDE process model.

Figure 8-3. Five Degrees of Delegation

HIGHEST DEGREE OF DELEGATION

5 **Teammate acts on own.** Task is teammate's to perform based a high-level of wisdom, experience, and judgment. Boundaries are most likely well known but still should be mutually confirmed. Formal review sessions minimal here; status through normal monthly reports or informal discussions will usually suffice. *ABCDE* model used as needed.

4 **Teammate acts on own, but full delegation process is in play.** Task is teammate's to handle, but within preset boundaries thoroughly discussed and agreed to between leader and teammate. The number of checkpoint reviews and their format is determined in advance to lock in the leader's role as helper. *ABCDE* model is utilized in full.

3 **Teammate acts, but only after obtaining leader's approval.** Task is teammate's to explore and prepare for action. Before proceeding, the teammate must inform the leader of the structure and process to be used to accomplish the task objectives. It will be teammate's to do after approval. *ABCDE* model is fully employed.

2 **Teammate is provided with a set of predetermined alternatives.** Task will be the teammate's to do after analyzing a limited number of options provided by the leader. The teammate chooses an option for action; the leader accepts the final choice and institutes the full *ABCDE* model with the teammate in carrying out the chosen alternative.

1 **Teammate is asked only for analysis and recommendations.** Task is to assess the problem/situation and present the facts and reasoned thinking to leader. Leader decides what to do. At the leader's discretion, the teammate may be delegated simple assignments with restrictive boundaries as a learning experience under close control of the leader. *ABCDE* model fully employed in those instances.

LOWEST DEGREE OF DELEGATION

IN NO CASE DOES DELEGATION INVOLVE DUMPING

Effective delegation to increase one's capacity to act is an especially valuable course of action anytime (1) you significantly add more responsibility to a task your teammate is currently performing, (2) assign responsibility for an entirely new project to a person, or (3) promote an individual to a new position containing any number of new duties and responsibilities. In any instance where responsibilities are enlarged, the commensurate authority must be delegated and accountability obligations must be understood.

Delegation does not apply to the teammate's normal, ongoing duties and responsibilities, which are well in hand and performed on a routine basis. Although any major changes that alter the routine and increase responsibility are a call for delegation. *However, at no point does the ABCDE model or the degrees of delegation scale suggest dumping!*

"Dumping" is hit and run. It is tossing a bunch of tasks on another teammate—often routine, repetitive, and undesirable—and telling them to handle those chores with little or no explanation of priorities or clues about what needs to be done. Dumping is giving the teammate more work to do, which increases their workload, stress, and pressure without serving any developmental or growth purpose. And this extra workload, which has the teammate working long hours and weekends, interferes with their ability to take on more responsibility. Anyone can dump the rubbish jobs; increasing capacity to act is about ensuring that the tasks and jobs you delegate create value for those doing them.

WHAT TO DELEGATE AND WHAT NOT TO

How are you to know what to delegate? This will vary, depending in part on the task itself, the current level of knowledge, experience, skill, and readiness of the teammate relative to the task in question; depending too on the purpose behind your delegation. The best rule of thumb is: you cannot delegate any tasks that are central to your role as manager—that is, those specific managerial duties only you can perform. Beyond those, you will find many terrific delegation opportunities of which you can take advantage.

General Opportunities for Delegation

- Delegate tasks that can be done better by a teammate because of his expertise or because the person is closer to the action and possesses more timely information.
- Delegate tasks that are relevant to the teammate's career development.
- Delegate tasks that are urgent but not high priority. Often this is either humdrum, required reports called "system satisfiers," or tasks no one wants to do—like picnic planning, canned food drives, or United Way collections. Spread these requests around to all teammates to avoid becoming a dumper. Every unit has mundane tasks to be completed, and all teammates should get their fair share of this action.
- Delegate tasks of appropriate difficulty to the teammate—challenging enough to stretch her, but not so difficult there is little hope of successful completion. (See Figure 8-3, Five Degrees of Delegation.)
- Delegate both pleasant and unpleasant tasks across the board. Don't hoard all the visible or fun tasks for yourself.

Tasks Central to Your Managerial Role Cannot Be Delegated

You may want or need to seek counsel from teammates regarding any of the managerial functions shown here, but the final decision authority always rests with you because all of the following tasks are central to your role as manager:

- Creating annual plans that fit within the larger strategies and goals of the business
- Setting goals and priorities for your work team

- Allocating resources (money, headcount, equipment, etc.) among your teammates
- Evaluating the performance of your teammates
- Making merit pay and promotional decisions regarding teammates
- Disciplining teammates regarding policy, ethics, and moral violations
- Directing teammates during crisis or legal investigation
- Supervising an assignment your manager has directed you manage

BENEFITS OF EXCELLENT DELEGATION

Skillful delegation employing the ABCDE model provides significant benefits to you as a leader, to your teammates receiving delegated tasks, and to the total organization in its effort to build a solid base of future leaders. Delegation's benefits are shown in Table 8-1.

Table 8-1. Five Benefits of Delegation

Benefits	Elaboration
Time	*Delegation increases discretionary time for you to complete more planning and strategic work than could be accomplished otherwise.* A word of caution, however! Delegate meaningful tasks when you are overloaded. Do not attempt to "dump" all your undesirable tasks to free up needed time.
Development	*Delegation is a primary method for developing your teammate's knowledge and capabilities.* New managerial responsibilities or technical skills needed in a higher position can be assigned to a team-

mate to develop the skills necessary to perform the higher-order responsibilities. Any time you appeal to areas of interest with appropriate built-in challenges, and use the ABCDE model, you will simultaneously develop and motivate your teammates.

Commitment

Delegation boosts your teammate's commitment to decision execution. Commitment will be stronger when a decision is made by your teammate; your teammate will feel ownership of the decision and seek to avoid an unsuccessful implementation that reflects poorly on his or her competence.

Decision Quality

Delegation improves decision quality with better efficiency, information, and timeliness. Delegation is more likely to improve decision quality if your teammate has more expertise in task performance than you do, or if your teammate's job requires quick responses to a changing situation and the lines of communication do not permit you to closely monitor the situation and make rapid adjustments. Since the teammate is closer to the problem, has more relevant information than you, and knows the boundaries, he or she can make quicker and better decisions.

Trust

Delegation improves mutual trust and confidence between you and your teammate. Your delegation of appropriate assignments to your teammate demonstrates your trust in that person. You're sending a message that says, "I believe you have the potential to do more, to expand your capacity to act." Your teammate values the trust and works hard to keep it. Research demonstrates that individuals who felt trusted by their leaders were notably more confident and effective than those who did not feel that way.

BARRIERS TO DELEGATION

Many leaders wish to transfer work to teammates for the reasons shown above, yet those same leaders *fail to let go so others can get going*! Why is this so? The causes of this paradox lie within both the leader *and* the teammate.

From the Leader's Perspective, Fear Is the Barrier

The major obstacle to successful delegation is an inherent human emotion: *fear*. The leader may be fearful that a teammate will not be able to perform the activity up to the required standard. Therefore the leader is afraid of being on the receiving end of negative consequences from her own manager should there be any displeasure with the results. On the other hand, an insecure leader may be fearful that the teammate will do a better job on the assignment than she would have, thus threatening her own job security.

Another dimension of fear from the leader's end may be the fear of ambiguity, of not being continually on top of every detail. This is also known as the "fear of losing control." Any leader takes a calculated risk whenever an assignment is delegated to a teammate. Even with clear instructions, dependable teammates, and mutually agreed-upon controls, the possibility remains that something can go wrong. However, unless the leader adjusts emotionally—as well as intellectually—to this element of risk, there is likely to be a reluctance to delegate anything to anyone.

Finally, a leader may be a risk avoider, and be afraid of delegating in any situation where very tight goals, standards, and timetables cannot be put in place to stay on top of every detail and track every action.

Regarding fear, Alfred P. Sloan, the highly respected former president and chairman of GM, 1923–1937, put it is as simply and pointedly as anyone: "Fear is understandable, but a manager who cannot stand the anxiety of delegation is fleeing from his managerial role. He ought to look for other work."

From the Teammate's Perspective, Risk Avoidance, Lack of Confidence, and Fear of Repercussions Are the Barriers

Inadequate delegation is not always due to the manager's fear. Teammates also can be at fault by resisting the leader's attempts to delegate.

A teammate may find it easier and less risky to ask the leader to decide. Making a wise decision is hard mental work. If a teammate knows any troublesome problem can be taken to the leader for an answer, it is natural to do so. In addition to being easier for the teammate, this course of action has another advantage: If the leader makes the decision, the teammate is less likely to suffer severe criticism from the leader for any negative consequences at a later time.

Sometimes a teammate may lack self-confidence in accepting delegation beyond level one or two in our *degree of delegation* scale. Ordering a teammate to be self-confident will have little effect. In many cases, however, a teammate will grow and develop self-confidence if the leader helps him realize his own potentialities by providing carefully chosen experiences with increasingly difficult requirements.

Being on the receiving end of needless criticism if things do turn out badly can cause a teammate to be less than enthusiastic about delegation. A teammate typically welcomes constructive reviews but resents unwarranted or condescending criticism. If past experiences from assuming greater delegation have commonly earned a blast of embarrassing or unwarranted reprimands for any mistakes or shortfalls, then a teammate naturally will be inclined to be cautious and play it safe. In other cases, a teammate may simply dislike the leader. For whatever reasons, the two individuals can't get along. The resultant attitude of the teammate is: "Why should I stick my neck out for this jerk?"

FIVE QUICK SENSING QUESTIONS FOR GAUGING ONGOING DELEGATION

As a collaborative leader, increasing capacity to act is an ongoing partnership-building process with individual teammates and small groups. In formal one-on-one discussions, as part of performance reviews, as a periodic agenda item at staff meetings, at a one-hour "team timeout," or any other appropriate one-on-one or team opportunity, you can gain an ongoing sense of how you're doing in this vital leadership dimension. I've had managers use these questions with countless teams. While the questions are simple, the feedback you get is powerful; it keeps your team on the cutting edge of success.

1. How am I doing in consistently applying the A+BC+D+E model with you?
2. In which of the four areas could we improve? What do we need to do differently?
3. Do you have enough of the right information needed to do your job?
4. Are you satisfied with the degree of influence you have in decisions that affect your work?
5. Am I available, when needed, to coach and counsel you regarding ongoing or new projects and/or assignments?

Keep in mind, asking these questions of an individual or a team implies that you are interested in collaborating to make changes wherever possible to improve any identified problem areas. Don't ask them if you have no intention of forming a collaborative partnership with the appropriate parties to make the situation better.

DELEGATION EFFECTIVENESS: KEY LEARNING POINTS AND WHAT I WANT TO IMPROVE DIFFERENTLY

My Key Learning Points from Chapter Eight:

What I Want to Do Differently:

READER REFLECTIONS AND APPLICATION ACTIVITIES FOR CHAPTER EIGHT

Reflections

To determine if a task you are considering to delegate is appropriate in a given situation, reflect on these five key questions:

1. Is there someone else who has (or can be taught) the necessary information or expertise to complete this task? Essentially is this a task that someone else can do, or at least can partner with me and learn to do?
2. Does this task provide an opportunity to grow and develop another teammate's skills? A big plus.
3. Is this a task that will recur, in a similar form, in the future? Recurring tasks are prime candidates for delegation.
4. Do I have enough time to delegate the job effectively? If time is not available for adequate training, for questions and answers, for opportunities to check progress, and for any necessary rework, the teaching aspect of delegation will be compromised. Still, don't use lack of time as a universal excuse not to delegate: carve out the time.
5. Is this a task where I can begin to bring in a teammate, under my wing, to begin learning a key business skill? Tasks critical for long-term success—for example developing operating plans, setting strategy, reorganizing your unit, recruiting the right people for your team—genuinely do need your attention with your taking the lead and acting as final decision maker; but do not eliminate bringing others in to work with you and learn. You can involve others in the process by delegating some of the "doing" pieces using the A+BC+D+E model.

If you can answer "Yes" to these five questions, there is a high probability that the contemplated task is worthy of delegation.

Leader Application Activity: Delegation Effectiveness with an Individual (30 to 45 minutes)

1. Have teammate read Chapter Eight and come prepared with answers to two questions in advance of meeting with you:

- What did you feel were the most important learning points from the chapter?
- Why were these key learning points for you?

2. Open the session by letting the teammate share his or her views on the two prework questions above. Actively listen, understand.
3. Share what you found important in the chapter and why these points were key for you.
4. Have the teammate present a past work situation where a task was delegated. Examine results in terms of each of the segments in the ABCDE model. Could employing the model have improved the process or results? Which parts would have helped? How so?
5. Have teammate present a current situation where task delegation may be an option. Mutually review the five questions in "Reflections" to determine if task delegation is appropriate. If so, mutually develop a delegation strategy using the ABCDE model.
6. Hold a postmeeting follow-up. Was the delegation experience successful? Regardless, if yes or no, what lessons were learned?

Leader Application Activity: Delegation Effectiveness with Your Team (45 minutes)

1. Have all teammates read Chapter Eight and come prepared with answers to two questions in advance of the meeting:

   - What did you feel were the most important learning points from the chapter?
   - Why were these key learning points for you?

2. Open the session by asking teammates to share their views on the two prework questions. Actively listen and understand.
3. Share what you found important in the chapter and why these points were key for you.
4. Discuss how excellent delegation could improve learning and task performance in your unit:

   - Give specific examples where delegated tasks and the results were below expectations. What could we have done better? Give specific examples where delegated tasks and the results met or exceeded expectations. What do we want to reinforce?

- Have we underutilized excellent delegation on our team? Is it a process we should use more routinely? Why or why not?
- Using Figure 8-3, "Five Degrees of Delegation," what opportunities exist at each of the five levels?

NOTES

1. R. E. Quinn, S. R. Faerman, M. P. Thompson, and M. R. McGrath, *Becoming a Master Manager: A Competency Framework,* 3rd ed. (John Wiley & Sons, Inc, 2003), 48–49.
2. L. Urwick, *Elements of Administration* (New York: Harper and Brothers, 1944), 51.

CHAPTER NINE

TEAM PROBLEM SOLVING I

A Systematic, Collaborative Model for All Occasions

CHAPTER OBJECTIVES

- To present a practical six-step team approach to identifying, analyzing, and solving problems in a systematic, collaborative manner
- To contrast the differences between problem sensing and problem definition
- To demonstrate how to write a problem definition as a "gap closing" statement
- To provide the how-to's for facilitating each step of this six-step model

INTRODUCTION

Sally, the wife of a busy executive, complained to her friend Clare that she never saw her husband because he came home late every night, and she was bored staying home alone. Clare chimed in immediately and said, "Oh, poo, I have the perfect solution to your problem. You just need to get out and exercise a little. It'll keep you busy and relax you too. Here, take my bicycle and ride it 10 miles every evening this week. You'll feel a whole lot better. Call me next week and tell me how you feel."

"Well, I don't know. Seven days of bicycling isn't something—"

"Sally, trust me. Do it! Now get along."

The next week Clare got a call from Sally. Clare excitedly asked, "Well, how do you feel now?"

Sally said, "Lousy, terrible, worse than ever."

"My word," replied Clare. "Why?"

Sally, in a sobbing voice filled with exhaustion, said, "I'm tired, sore, hungry, and I'm 70 miles from home!"

I'm sure those weren't the results Clare expected. But if you look back at the dynamics of the problem solving process that was used, then the results, while still humorous, aren't really surprising at all.

Far too many individuals and groups try to solve problems in a random, undisciplined, quick-and-dirty manner. A problem situation arises, a rambling, often argumentative discussion ensues, a clear problem definition is ignored, analysis is bypassed altogether, solution generation and selection turn out to be whatever the manager tells the group to do—or whatever the dominant coalition can once again railroad the group into doing. There is no time for discussion of the pros and cons of the solution or for a critical look at potential implementation problems. A solution is forced upon the group, and implemented by a group of noncommitted people. The results in most cases are disastrous. Not only isn't the problem solved, but matters are made worse. This, in turn, triggers a crisis, more pressure, and a more intense cycle of what happened previously.

The opening story, while involving only two people, illustrates the point about poor problem solving practices. You, as a collaborative leader, can do much better by utilizing and facilitating the systematic, collaborative problem-solving model presented in this chapter.

AN ORIENTATION TO SYSTEMATIC, COLLABORATIVE PROBLEM SOLVING

If you examine the behavioral science literature, you'll discover the actual steps employed in a wide variety of problem solving models fall within the range of four to nine steps. The basic processes are the same: The models with fewer steps tend to consolidate several of the discrete steps included in the longer models.

The six-step model, shown in Figure 9-1, was first developed and used as part of Xerox's corporatewide Total Quality initiative in the late 1980s and mid-1990s. I was privileged to be part of the team that helped refine and spread it across the company. It is a proven process that has been used tens of thousands of times around the world at Xerox. It has been liberally shared with other companies, where the model has become part of their culture.

The steps involved in our model form a generic process, and that is its strength. This model is applicable to any kind of problem solving situation, whether it occurs in your head, in a two-person group, in a committee, or in the total unit. I have expanded the original wheel a bit by adding the fuzzy problem-sensing phase.

A generic, structured process provides a method that you and your teammates can understand and use to greatly improve the synergy among yourselves. In any meeting, at any time, at any location, if a teammate says, "I propose we use our problem solving model to tackle this one," everyone around the table will understand exactly what steps and disciplines are required.

Although the six steps are shown in Figure 9-1 as a wheel and numbered sequentially, a group seldom glides smoothly from Step 1 through to Step 6 without having to make several loops back to revisit and revise information from earlier steps. The realities of problem solving are such that it often proceeds by fits and starts, both rationally and emotionally; however, the total process does proceed through each of the six steps identified in our model.

By moving a team around the wheel, you actually will be facilitating the group through a series of divergent and convergent thinking activities. Divergent thinking is used at idea-generating stages—like Steps 2 and 3, when the group is exploring the differences and creativity among members. Convergent thinking is used at idea-sorting and idea-selecting steps—like Step 4, when the group needs to evaluate solutions and agree on the best one(s) for implementation.

Figure 9-1. The Six-Step Collaborative Problem Solving Model

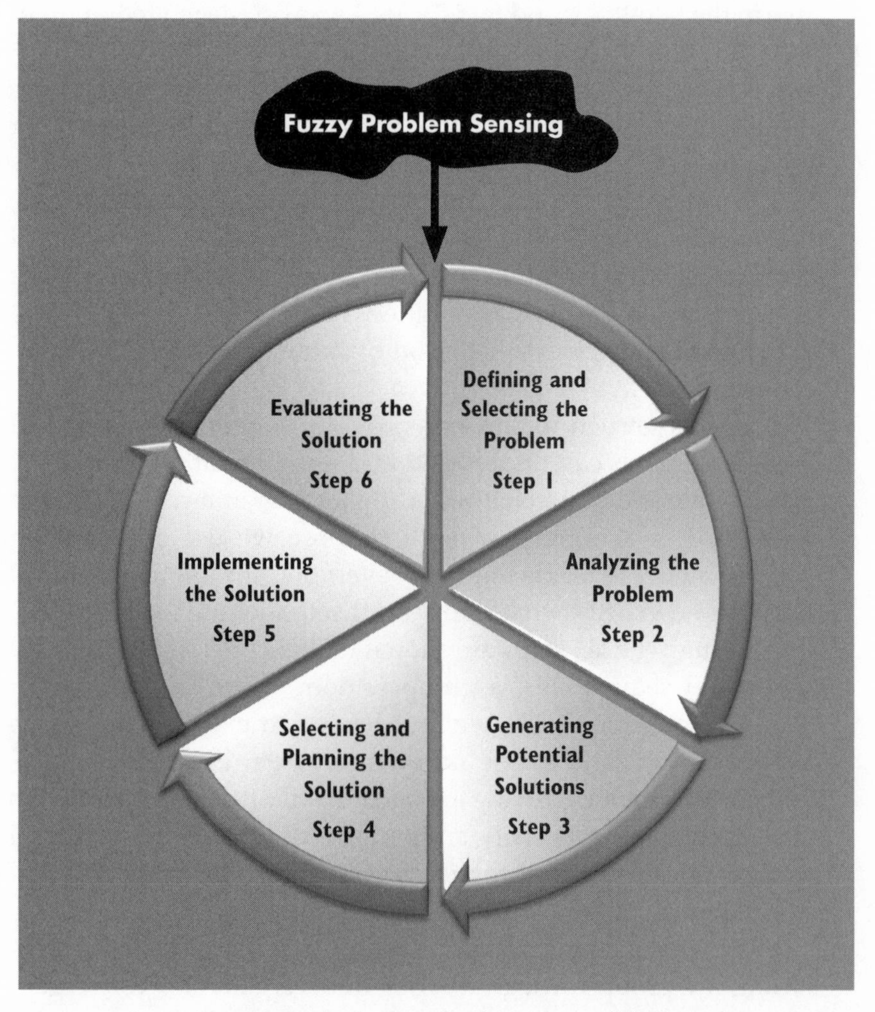

Source: The six-step wheel reprinted with permission of the Xerox Corporation

It is best to think of this problem solving method as your guide, or a kind of road map, for facilitating collaborative problem solving efforts. This process provides structure and direction to your facilitation activity. It points the way for addressing difficult problems in a thorough, step-by-step manner. The more complex a problem, the more useful a systematic, collaborative process becomes. The specifics of the full process will be covered shortly.

Pitfalls Avoided

The collaborative model advocated here helps you and your team avoid six notorious pitfalls of ineffective problem solving:

1. Failing to gather critical data about either the problem or proposed solutions.
2. Jumping to a solution before effectively analyzing the true nature and dynamics of the problem under consideration.
3. Tackling problems that are beyond the control or influence of group members.
4. Working on problems that are too general, too large, or not well defined.
5. Allowing habit and past experience to restrict mental flexibility in generating and evaluating potential solutions.
6. Failing to adequately plan how to execute and evaluate recommended solutions.

Facilitation Benefits Gained

The model makes you a better collaborative leader by expanding your skills in five critical ways.

1. It gives you a common process and language that you can teach all teammates to use.
2. It arranges your team's problem solving activities in a logical, sequential order so everyone is following the same steps in the same order.
3. It focuses everyone's attention on what needs to be done at each step of the process.

4. It helps you and your teammates recognize the potential pitfalls mentioned earlier by asking critical questions at each step of the problem solving process.
5. It helps you and your teammates decide the right time to move on to the next step.

PROBLEM SENSING: PRECURSOR TO PROBLEM SOLVING

Closely allied with problem solving—but antecedent to it—is problem sensing. Problems typically do not come to a group neatly packaged with a big bow and a flashing neon sign reading PROBLEM. Most of the time the true problem is either unknown, disguised, ill-defined, buried in a mass of data, or some combination of these conditions. All that exists is a feeling held by an individual or group of individuals that some state of affairs is unsatisfactory.

William Dyer clarifies the problem-sensing cycle that is often a precursor to the problem solving process:

> [In sensing,] the manager begins not with a problem but with a concern or feeling of unease that perhaps there are areas that should be improved. With this concern as the motivating influence, the manager begins to gather data by talking to people, asking questions, perhaps initiating a questionnaire or survey. The focus of all this is to find out if there are any conditions that are keeping people from being as effective as they could be in their jobs. Having gathered data, the manager then analyzes the information to see if there is enough evidence to decide that there is, in fact, a problem or problems that should be solved.[1]

Problem sensing is no small task because the problems you face—regardless of your role, responsibilities, and formal position in the organization—are actually clusters of information and observations from which meaning must be extracted. The volume, variety, and constant disruptive elements in the stream of information you receive each day make problem sensing an activity worthy of your concentration.

You alone—or in conjunction with several others—processing myriad data and information streams received over time, begin to feel discontent with particular aspects of how your total organization, department, bureau, agency, or work team is performing. Your feeling of apprehension that something is amiss is your internal signal that problem sensing is taking place.

Four Examples of Problem Sensing

- "I don't know what the problem is; all I know is our inventory control system is broken. We don't have any idea what we have in our warehouse or where to find anything."
- "Our billing methods are all screwed up. We have got to do something. Patients are being billed two and three times for lab tests and medicines they have already paid for."
- "The new reporting system for measuring store performance just isn't working; the whole thing is a disaster. No one can make heads or tails out of these reports."
- "Look, our costs are way out of line. Our competitors can sell their services and make a tidy profit for what it costs us just to provide those same services with no profit. That's outrageous."

In all four examples cited in the "Problem Sensing" sidebar, the red light is blinking. Problem awareness, based on symptoms, is being noted. *Fuzzy problem sensing is taking place, not problem definition.* As you will see in Step 1, correct problem definition never involves symptoms.

Many groups get stuck at this point. They feel strongly motivated by some mess that is causing fuzzy problem sensing to occur, but no one on the team—including the manager or chairperson—knows how to move forward; yet the felt need remains strong. Without a systematic process capable of producing meaningful action to remove it, the group flails and struggles, with nothing to show for its efforts except an abundance of frustration, anger, and a cynical "I don't give a damn" outlook toward future group problem solving initiatives.

If the group does decide to tackle the situation by plunging ahead, it does so in the worst possible way. It takes the fuzzy problem sensing and the vague statement about it by someone as *the* defined problem, skips the analysis, jumps immediately to concocting an ill-defined—and often incorrect—solution, implements the solution, then discovers that not only didn't the solution work, it caused other problems that made matters worse.

Fuzzy problem sensing does perform one very important function: it sets the stage for problem solving. During problem sensing you should be making a determination of whether a problem exists that is worth pursuing through utilization of our six-step collaborative process. And if the answer is yes, you enter the collaborative model at Step 1 and go from there.

STEP 1: DEFINING AND SELECTING THE PROBLEM

The output of this step is the correct definition and selection of a problem for further problem solving action. This is the most important and possibly the most difficult step in the process. Poor facilitation here leads to the deadly trap of identifying and defining symptoms as the real problem. This in turn leads to the development of a solution for the wrong problem. There's an old saying: "Nothing is as useless as the right answer to the wrong question." Time spent in attempting to isolate the real problem, therefore, is time well spent.

Remember where you are at this point. A vague problem sensing statement that identifies a fuzzy problem area may have been communicated to you by a higher authority, and now you've pulled together a cross-functional problem solving group to begin defining and tackling the problem. Or you might have been motivated by feelings of discomfort when a fuzzy, ill-defined problem began to cause trouble for your work unit. Knowing that this fuzzy problem is worthy of attention because it is the sign of a deeper issue, you in collaboration with a small group of hand-picked teammates will now assume the task of resolving it. Finally, the fuzzy problem might have been mutually sensed by you and your direct reports, and now, together, that same group will attempt to solve the problem.

It is important to keep in mind that the fuzzy problem may arrive at your doorstep in a variety of ways. However, do not be fooled into believing you have identified a problem; at this stage you haven't. Still, how you facilitate the group from this point forward will be critical to the overall success of your problem resolution effort.

Perceive Problems as Gaps

The most vivid and accurate way to understand a problem is to view it as a gap between what exists and what is desired.

An excellent working definition is: A problem exists anytime there is a difference between a current state (as is) and a desired future state (what should be).

For example, a problem exists when our customers are buying only 1,000 units a month and our strategic plan forecasted 2,000 purchases a month. Or, another example, a problem exists when you and your family are in London's Heathrow Airport without your luggage, when you and your family should be in Heathrow with your luggage.

Writing an excellent problem statement means first defining the two conditions: the "as is" condition, which describes the situation as it currently exists, and the "desired future state" condition, which describes the result if the problem is successfully solved. Finally, the problem statement itself is developed. This is always written in terms of: *How can we close the gap between the current and the desired state?* The following example will illustrate the problem definition process.

Current state: On average, 15 percent of all sports cards produced each month are miscut off-center.

Desired state: Beginning with this year's new line of football cards, on average, only 0.5 percent of all sports cards produced each month are miscut off-center.

Problem statement: How can we *close the gap between our current state* of having, on average, 15 percent of all sports cards produced each month miscut off-center *and our desired future state* where, beginning with our new football cards, only 0.5 percent of all sports cards produced each month are miscut.

Facilitating Development of the Current Situation Statement

The person, or small group, sensing a problem believed worthy of more scrutiny has to put in some organized thinking to craft a concise current situation statement.

There are three guidelines for helping you write the *current state* piece for any problem statement:

1. *Do specify the extent.* Our "as is" description clearly specifies the extent of the current situation—on average, 15 percent of all sports cards we produce each month are miscut off-center.
2. *Do* not *specify causes.* Our "as is" statement does not specify that "90 percent of all miscuts are made by people with less than six months' service working on the manual cutting machines." We don't know that yet.
3. *Do* not *specify solutions.* Our "as is" statement does not specify that "we need to design a one-day skills training program for all cutting-machine operators and to reset the operator skill mix on all shifts." We don't know the best solution yet.

Our current state statement is well written. It makes no assumptions about who or what caused the miscuts, what should be done about the miscuts, or how terrible it is that miscuts exist. The statement is a clean, succinct description of a current situation. As Officer Joe Friday would say each week on the classic TV police show *Dragnet*, "Just give me the facts, ma'am."

The Desired State

This is a succinct description of what the future would look like if the current state is resolved. It is like a lighthouse beacon providing constant direction throughout the gap-closing transition. Your desired state should be written as objectively as possible so it can serve as a meaningful benchmark against which to evaluate the effectiveness of the solution implementation. In our example we will know we've solved the problem if, beginning with the new line of football cards, on average only 0.5 percent of all cards produced each month are miscut.

The Problem Statement Itself

While this statement is straightforward, it must be written in terms of "gap closing," since closing the gap between the current and the desired state is in fact the problem facing the group.

> An incisive problem statement easily can be written by following the sentence structure: "How can we close the gap between our current state of . . . and our desired state, where . . ."

Getting Away from Defining Problems in Terms of Symptoms

By defining problems following the method outlined here, you virtually eliminate the single most common problem solving mistake: defining the problem in terms of symptoms. Symptoms are the surface side effects of the real problem. The true problem usually is buried underneath a pile of symptoms. It is a grave mistake to attempt to define the problem in relation to your first dissatisfactions. The reason is that these early dissatisfactions typically are reactions to the symptoms of an underlying problem, not to the problem itself.

In summary, let your initial dissatisfactions and your fuzzy problem sensing simply be the lead-in to Step 1 of your problem solving effort.

By systematically developing a succinct situational statement of the current condition (free from expressed or implied causes and solutions), paired with an objective statement of the desired future condition (the result if the problem is successfully solved), you've set up your problem as a gap-closing activity. Then, by writing your problem statement so it is focused on *closing the gap* between the current state and your desired state, you've eliminated the serious error of defining the problem in terms of symptoms.

Edgar Schein, writing on this subject, provides insight to another dimension of this situation. Defining the problem in terms of symptoms invariably leads the group to skip over the analysis phase and jump directly to generating solutions:

> Let's take the example of sales falling off [which is a symptom]. . . . Manager X has called together his key subordinates and they sit down to discuss "the problem" of declining sales. If the manager is not sensitive to the issue [of proper problem definition], he may soon be in the midst of a discussion of whether the advertising budget should be raised or 10 more people should be sent into the field. But has he as yet defined the problem? . . . He doesn't know what he really should be working on.[2]

Poorly Written Problem Statements Revised

Each of the following examples shows weak aspects of problem definition, the reasons a particular segment is weak, and a proper restatement of the problem.

Example 1

Weak problem statement: the time to perform the lube-oil-filter services at our three service centers is too long.

This is not specific enough. How long is "too long"? A proper statement would be:

Current state: 58 percent of all lube-oil-filter services in our three service centers take more than 20 minutes.

Desired state: Response time for every customer's lube-oil-filter service is 20 minutes or less.

Problem statement: How can we *close the gap between our current state* of having 58 percent of our customers' LOF service taking more than 20 minutes *and our desired state* where every customer's LOF service is completed in 20 minutes or less?

Example 2

Weak problem statement: People are not contributing as much to the United Way Campaign this year because the economy is in a recession.

This is a cause. An economic recession is only one among many possible reasons for diminished United Way contributions this year—it's not the problem. Also, this statement is not specific as to how much campaign contributions are down compared to last year. A proper statement would be:

Current state: With two months remaining in our annual campaign, the dollar amount of United Way contributions is down 20 percent versus last year.

Desired state: By the end of our campaign on April 30, United Way contributions will exceed last year's dollar amount by 10 percent.

Problem statement: How can we *close the gap between our current state* of United Way dollar contributions running 20 percent below last year, with two months remaining in our fund drive, *and our desired state* of exceeding last year's dollar amount by 10 percent at the campaign's close on April 30?

Example 3

Weak problem statement: We need to change the set of universities we are currently using to recruit our new college hires.

This isn't a problem, it's a solution. What's the problem? It might lead to an entirely different solution. A proper restatement would be:

Current state: 35 percent of our new college hires leave the company within 18 months of being hired.

Desired state: Retain 85 percent of our new college hires for at least 5 years.

As you can see, both the current- and the desired-state elements are critical to excellent problem definition. Notice too, besides the weaknesses identified in each of the original examples, none considered the desired future state—another common weakness. Often you and your group will discover that you do not have enough information to create a proper problem definition. If that is the case, move on to Step 2 and collect data about the extent and nature of the current-state situation. Then, with specific information in hand, move back to Step 1 and refine the problem statement.

The "as is" element of the problem statement is the most difficult piece to define since it represents a condition that is going on right now. It is dynamic, it is happening, it is real! The desired-state part of the problem is merely a condition you would like to achieve in the future, and as such, it is easier to develop. The focus for Step 2 always will be on analyzing and fixing the here and now—the current situation.

Finally, in addition to properly writing an excellent problem statement, if you put the words into a pictorial framework, you can increase its communicative impact manifold. Therefore, my recommendation is that you use the visual format shown in Figure 9-2 for all your problem statements. It is clear, it graphically reinforces the gap challenge, and it depicts the problem definition in a manner everyone can easily understand.

Figure 9-2. Visual Format for Writing Excellent Problem Statements

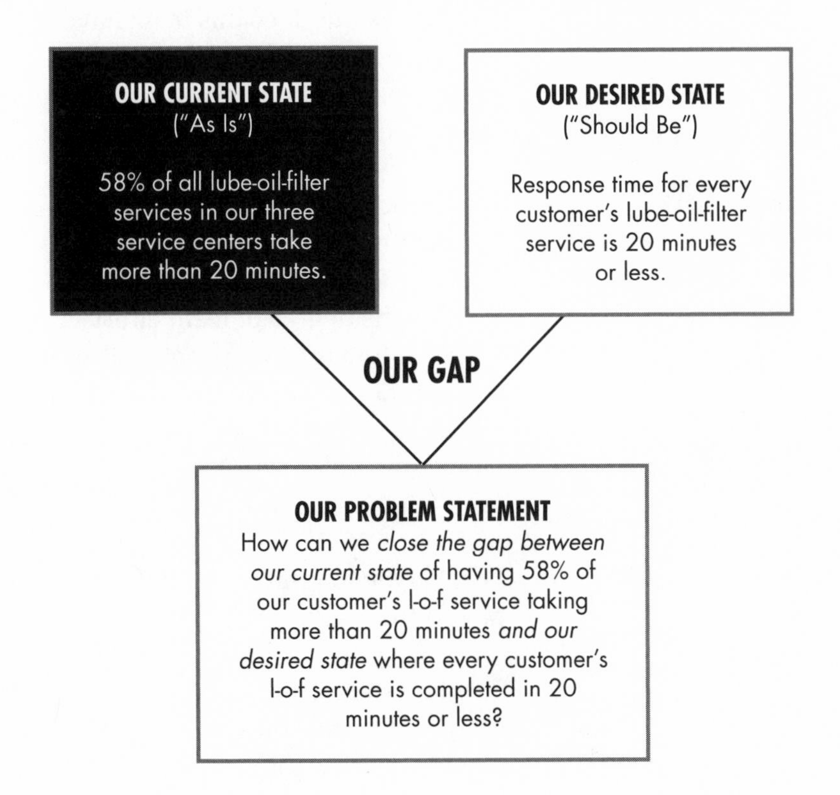

Selecting a Problem

Sometimes you and the group may have identified and written several different potential problems to work on in the course of working through this first step, but those involved know the team does not have the process capability to work on all of them. Wearing your facilitator's hat, your job is one of leading a whole group discussion to select the problem the team will work on, or the one it will work on first. I have found that focusing your discussion around the six criteria shown in Table 9-1 will be an immense help in selecting a problem if you face several choices.

Table 9-1. Problem Selection Criteria

Control	To what extent does this team have the influence and/or authority to bring about closing a gap between the current state and the desired future state?
Importance	How much does it matter whether this problem is solved?
Difficulty	What is the degree of team effort required to work this problem through to a solution?
Time	How long will it take to resolve this problem?
Return on Investment	What is the magnitude of the expected payoff from solving the problem?
Resources	To what extent are the necessary resources to solve the problem accessible to the team (people, money, equipment, etc.)?

You and your team may want to add or substitute other criteria. This is fine. However, if you do include any other criteria, make certain everyone agrees with and understands them. The ones shown here are guidelines. In the next chapter we'll see how these guidelines can be incorporated into a criteria matrix to quantify problem selection.

As a facilitative leader, don't allow the group to get bogged down in nitpicking detail or into a destructive argument over applying precise measurements to the criteria if you are using them as discussion guidelines. It is no crime to move on to Step 2 and do some analysis of a chosen problem and discover that you need to return to Step 1 and either redefine the problem or choose another one to work on. Hang loose; be flexible.

STEP 2: ANALYZING THE PROBLEM

After a problem has been defined and selected for further work, the next phase to be facilitated is analysis. The most important thing to keep in mind here is that the focus of your collaborative leadership effort always is on analyzing the current state to uncover the cause or causes for what exists. With this understanding, you then will be able to look for ways to eliminate the discrepancy between what exists and what is desired.

For some problems what we know and/or what others know may provide more than enough information to do a superb job of analyzing the problem. The gold nuggets of wisdom accumulated over the years in the heads of people are a powerful source of information and understanding. However, often personal knowledge is not enough. It can be subjective, slanted, and value laden. We need factual data to supplement our personal knowledge.

Three words best describe the group effort during the analysis phase: dig, dig, dig. Giving and seeking information and opinions, proposing, and building on proposals and ideas, are important interpersonal skills to emphasize at this time. Feelings and emotions can be quite strong and they must be facilitated. Facts will be critical to developing a meaningful analysis; they will often be intertwined with feelings. Facts need to be assembled and studied after people's feelings have been heard, understood, and accepted.

Some Fundamental Probing Questions

Rarely will you be able to assemble all of the information you would like to have to analyze the problem. Some data may be difficult or costly to secure. Time may limit the amount of information that can be gathered. As part of your planning process, you will help the problem solving effort if you think through what information is needed and in what priority.

The following nine questions, while short and to the point, are an effective "facilitation shovel" for the digging task that needs to occur to comprehend fully the current "as is" state.

1. What is happening with respect to the current situation?
2. Where and when is it occurring?
3. What evidence do we have of this? (For example, identify concrete instances or examples of events that support existence of the current situation.)
4. What are the dimensions (size, scope, severity) of the current situation?
5. What is affected? Who is affected?
6. When was the current situation first recognized as undesirable?
7. Is the current condition a unique or recurring situation?
8. What similar situations have occurred in the past?
9. How relevant are those past experiences to this current situation? (Did similar conditions, objectives, or ground rules apply then?)

Remember, at this point you and the team are digging, probing, clarifying; you are working together to get a better handle on the details behind the "as is" state. The old standby questions of who, what, where, when, and why, as well as how many, how big, how much, and the like, are invaluable at this point. Help the group identify cause-and-effect connections between events. While not a true brainstorming session, the atmosphere should be free, open, casual, and supportive. Get thoughts, ideas, and perspectives out of people's heads and recorded on flip chart paper.

A Process for Facilitating the Nine Questions

You can make this a very powerful and dynamic activity by having each of the previous nine questions written on a separate flip chart page and posted around the room prior to starting this activity. Then, taking each question in turn, you and your teammates give your views while the scribes write them down.

As an example, if someone makes a good point for flip chart page 8 ("Similar situations that occurred in the past"), while you are gathering information for flip chart page 1 ("What is happening with respect to the current situation"), have the scribe note that point on flip chart page 8. Then refocus attention on page 1 of the flip chart. Keep the information flowing, stay loose, and get everyone involved. Open the gate for quiet members.

If you or the group members want to rewrite, eliminate, or add questions to the suggested list covered here, do it. In fact, this modification activity is quite good for pulling everyone into the process and generating team ownership of the analysis. Ask what questions need to be answered to build a complete picture of the problem.

After the information is assembled on the flip charts, work with the group to organize it in a form that makes sense by probing with these two questions: What does it all mean? How does it all tie together?

Sometimes at this stage you may discover, based on the new inputs about the current state, that what was first defined as the problem is not as good a definition as first conceived. Other factors may have been uncovered that force a restatement of the problem. It may be necessary to return to the first step to redefine the gap in terms of the current and the desired state. This should be accepted as a normal aspect of effective group problem solving; it does not indicate failure on your part as facilitative leader.

Tools to Facilitate Problem Analysis

The flow chart, the cause-and-effect diagram, the force field diagram, and the Pareto chart, all of which will be detailed in Chapter Ten, are especially useful tools for Step 2. All four are simple to facilitate and provide a structured process for analyzing the problem. They are most effectively employed after you've held the "probing questions discussion" to help everyone understand and get within the same frame of reference regarding the "as is" condition. A brief overview of these four tools is included here.

- *Flow charts* are a way to diagram how a work process operates (flows) using a set of symbols and arrows. They are an excellent tool for isolating bottlenecks, unnecessary steps, unclear decision points, not enough decision points, too many movement steps, pressure points, and the like.

- *Cause-and-effect diagrams* (fish bones) are a systematic way of looking at a defined effect and the causes that contribute to that effect. Similar causes are grouped together for clarity.
- *Force field analysis* identifies those factors that both help and hinder closing the gap between the current state and the desired state.
- *Pareto charts* are an excellent way to rank order, by frequency of occurrence, the identified root causes of a problem situation. They focus the group on the significant few causes that are the biggest contributors to the problem.

If at some point your analysis gets bogged down, call a process check to see if missing information or data is causing the stagnation. Use the group to help identify what is missing and where the information or data might be located. If need be, stop your current session and appoint the appropriate people to go off and obtain what's required. Maybe a subject matter expert (a financial analyst, strategic planner, quality specialist, person from the local community, or the like) needs to be brought to the next session to share specialized information that you and the group require to properly analyze the problem. If so, identify who is needed and get them to the next meeting.

Two Deadly Traps: Nitpicking vs. Generic Fluff

Two common traps that ensnare and destroy groups during the analysis activity are the nitpicking trap and the generic fluff trap. With nitpicking, as the analysis unfolds, the group gets entangled in destructive arguments over irrelevant microdetails. At the other extreme, with generic fluff, the group analysis is so roundly worded and superficial that it is virtually useless.

Be attuned to these two traps. Even if you miss the signals, someone is sure to eventually call a process check and complain about the stalled analysis because: "We're sitting here fighting over the irrelevant nits," or "We're just wasting our time going through the motions of an analysis with our superficial, meaningless dialogue." Regardless of the source of the discovery, your task as facilitator is to move the group out of whichever trap it is in. You have two opposing strategies that you can employ to make progress—*generalizing* and *exemplifying.*

If the group is fixated on nitpicking detail, use the "generalizing" strategy. Get group members to broaden their thinking by saying things like, "The details of what a revised agencywide recognition event for superior performers would look like is not as important right now as understanding the pros and cons of whether we should even hold an event. Let's stick with this broader question for now." Or: "Fighting over the wording and format to improve our performance appraisal form isn't the issue we should be addressing now. Let's refocus on the more general question of root causes of employee dissatisfaction in the accounting department."

If the group is on a shallow, insubstantial, analytical journey, use the strategy of "exemplifying" to correct it. Bring the members down from the clouds and get them to be more incisive in their analysis by asking for specific examples. "What do you mean by 'funny things going on'? Give me some concrete examples of funny things." Or: "I keep hearing everyone talking about lack of help. Would each of you please describe a situation you personally were involved in that exemplifies a lack of help?"

Keep the analytical processing shifting as required from the general to the specific or vice versa to make certain a complete examination of the problem takes place.

STEP 3: GENERATING POTENTIAL SOLUTIONS

This is the fun part, the creative, imaginative, thinking-up stage of our collaborative problem solving process. Once the key cause or causes have been isolated, you and your teammates need to search for as many ways as possible to reduce or eliminate these root causes. Remember what was said earlier: "With a problem defined as a gap between what exists and what is desired, *problem solving involves the development of solutions to reduce or eliminate the causes of the defined gap so that this gap is closed."*

Since all identified root causes rarely can be resolved at once, attack the cause, or the significant few causes, accounting for most of the gap. A Pareto analysis often is useful at this point. The concepts of divergent and convergent thinking will now play a major role. Initially, the objective is to produce as many ways as possible—including some wild ideas—to solve the problem. Later these ideas will be screened and evaluated. Remember, it is far easier to tame a wild idea than to invigorate a timid one! Throughout the entire solution generation activity, support for ideas and for those who propose them needs to be high.

Tips for Generating Potential Solutions

The following tips will help you build and maintain collaboration within the team as you all work together to uncover potential solutions.

- *Review the results from Steps 1 and 2.* Before getting on with solution generation, review the problem statement from the first step and reexamine all the data collected and analyzed in the second problem solving step. A careful review of previous work often leads to ideas in the search for solutions.
- *Use brainstorming to get ideas for solutions.* Brainstorming will be covered in the next chapter. The freewheeling method is most often used at this juncture. Your main task as leader is to review the rules of brainstorming and then make certain teammates adhere to them. The scribe should record suggestions quickly and verbatim on a flip chart.
- *Draw from past experience.* If things get bogged down, ask teammates to compare this problem with a similar one from the past. What actions were taken? Which ones were effective? What might be increased, decreased, reversed, substituted, rearranged, combined, or adapted from the solutions previously generated to work here?
- *Bring in additional people for your group.* The solutions generation step is a fine opportunity to involve people outside your group. Coworkers and managers can bring insights

from their level in the organization. Resource people or functional experts can bring special expertise and additional perspectives when developing technical solutions.

- *Include and help your scribe.* First, remember your scribe is a team member and you must be alert to solicit her ideas throughout all team activities. Second, help her by testing for comprehension on technical or complex proposals to make sure they have been heard correctly. Also, for your scribe's benefit, repeat what a person says. That way she hears it twice and has an additional few seconds to write it down. Bring up a second scribe to another flip chart if the pace gets too fast and the original scribe can't keep up.
- *Keep encouraging everyone, and open the gate for quiet members.* Saying things like, "Okay, that's nine ideas, I'm sure we are not through yet," or "We're really cooking now; let's keep going," or "We're doing fine; let's see if we can't come up with five more ideas," all produce and maintain energy and momentum. Watch body language and who is not involved. Bring them in. "Art, you've always got good ideas. Toss one up on the flip chart." Or, "Carrie, I can see you have an idea on the tip of your tongue; what is it?"
- *When you finish, congratulate the group for its effort.* "Well done. Twenty solutions are a lot for a tough problem like this one. Now we've got the raw ideas to lick it." Generating a list of potential solutions takes time and energy; the group should be publicly recognized for its effort.

STEP 4: SELECTING AND PLANNING THE SOLUTION

Once the list of solutions is made, it is time to evaluate them, select the best solution or set of solutions, and plan for implementation. Combining some of the potential solutions into a set of solutions, mixing and matching for the best outcome should be done at this point. A basic set of evaluation criteria needs to be used to assure that each potential solution gets a fair hearing.

Going with what seems the most favored solution, or bowing to the preferred solution of the boss or the most vocal team member, is counterproductive.

Selecting the Solution

The question at the forefront of this half of Step 4 is: "What is the best way to close the gap between our current situation and our desired future state?"

Even though a particular solution may not work on its own, it may have elements that are good. Take time to combine the good parts of various ideas into new alternatives, then evaluate these critically. As in Step 1, a simple, generic set of criteria shown in Table 9-2 can be used as the evaluation filter. *A word of caution:* to avoid paralysis by analysis, don't agonize and nitpick over very small differences between options, particularly if those differences are smaller than the group's ability to evaluate reliably.

At this point your collaborative leadership task is one of leading a whole group discussion on the criteria to be used in selecting the solution. Use the suggested criteria set shown here as a starting point. While not exhaustive, this list has stood the test of time and experience as a creditable set for holding a penetrating discussion. These criteria will be quantified and folded into a solution prioritization criteria matrix in Chapter Ten.

Table 9-2. Solution Selection Criteria

Control	To what extent can implementation of this solution be managed or significantly influenced by the team?
Appropriateness	To what extent does this solution solve the problem by closing the gap between the current and desired state?
Acceptability	To what extent will the people impacted buy into and commit to the solution?

Time	How long will it take to implement this solution?
Return on Investment	What is the magnitude of the expected payoff from implementing this solution?
Resource Access	To what extent are the key resources (people, money, equipment, etc.) available for implementation?

Just as with the problem selection criteria discussed earlier in this chapter, you and your team might want to add or substitute other criteria. This is perfectly acceptable. If you do include any other criteria, make certain there is a consensus around both their definition and the desire to include them as part of the decision process.

The Table 9-2 list is a great foundation and can be used in most situations without any alterations. In any case, a simple set of criteria provides focus and structure to the solution selection discussion, which is fundamental to any problem solving process.

Planning the Solution for Implementation

Unless the selected solution is converted into action, it has very little value. The second half of Step 4, then, is to plan how the solution should be implemented and monitored. At this juncture you'll take your potential solution(s) and create action plans following the Action Planning Template shown in Figure 9-3 or a similar document of your choice.

Remember, for your implementation actions, the devil always lies in the details. Getting sloppy or lazy here will undo all your fine work up to this point. The following sequence of eight steps will guide you in creating a solid solution implementation plan.

1. Write the problem statement at the top of the action plan as a gap closure.
2. You may have several solutions for closing the gap. Each solution gets its own numbered plan but with the same

Figure 9-3. Action Planning Template Example

Your Problem Statement (Written as a Gap Closure):

Solution Action #______ to Help Close the Gap:

IMPLEMENTATION					TRACKING	
Major Activity Steps for This Solution to Close the Gap	Responsible Person	Target Dates		Resource Requirements	In-Process Measures	Final Success Measure
		Start	End			

problem written at the top. For example, if three implementation solutions are chosen to close the gap, there should be a set of three action plans to solve the defined problem.

3. Specify and clarify the major activity steps to be carried out for each solution—this includes obtaining any approvals required from others outside the team to move ahead.
4. Sequence the tasks—decide on the order in which they should be done.
5. Set responsibilities for the tasks and list name(s) beside each task.
6. Establish a schedule, with start-end dates for each task or activity.
7. Establish a tracking system—set up a simple monitoring system of in-process measures to track whether each task is being performed as planned, and set a final success measure for each task indicating its completion.
8. Use the desired future state of the problem statement as the constant overall goal against which the effectiveness of the whole effort is always measured.

STEP 5: IMPLEMENTING THE SOLUTION

This step consists of carrying out the action plan developed during the second phase of Step 4. To make sure the blueprint for implementation is adhered to, progress needs to be monitored. Checking the implementation as it unfolds requires you to be vigilant over three aspects of your plan: (1) each of the tasks involved in solution implementation, (2) each task's resource levels, and (3) the dates by which the tasks need to be completed.

Life in the real world, being as complex as it is, means sooner or later your plan most likely will need to be modified, and contingency plans set in motion, as unforeseen difficulties or opportunities appear. The experienced team expects this, does not get too upset when it happens, and gets on with whatever corrective actions are possible.

In a related point, Baldwin, Bommer, and Rubin, in their book *Developing Management Skills*, bring up an implementation fact worthy of note.

> While implementing a solution, many problem solvers unfortunately find they underestimated the problem's scope. . . . Although underestimating is discouraging, nothing is gained by staying the course simply to be perceived as consistent or confident in the solution. In the course of implementation, if you uncover significant information indicating [you are on the wrong track], stop the implementation. Many managers have been burned by implementing solutions they knew were incorrect but forging ahead even with this knowledge. Retreating so far along in the process will cause pain in the short term, but in the long term you will have acted appropriately.[3]

No matter what obstacles or opportunities arise, the better prepared you are with a well-laid-out plan to begin your implementation process, the easier it will be to make changes or develop contingency actions whenever necessary.

STEP 6: EVALUATING THE SOLUTION

Once the solution gets implemented, the team may think the problem solving job is finished. Not true. You and your team are responsible for obtaining direct feedback on the implementation outcome from the stakeholders in order to determine whether the proposed solution actually solved the problem.

The process is not complicated. As the leader, you need to help the team collect data on its results and compare this information with the desired-state portion of the problem statement created in Step 1. This comparison will indicate if you and your team have achieved your goal. In reality, most solutions are a mixed bag. They are neither tremendous flops nor stellar successes. Typically, some elements of the solution are implemented better than other aspects; this is to be expected. However, knowing the condition of the implementation effort at all times enables the team to make improvements or modifications that will help solve the problem.

If the desired state is not satisfactorily achieved, then you may need to begin the problem solving process over again (this is why it is drawn as a continuous wheel). Even if the desired state has been met, you need to help the team continue to monitor the situation to make sure (1) the solution continues to work and (2) no new problems are created by the solution itself. Finally, care must be taken in this step to avoid creating a bureaucratic nightmare or cumbersome systems for checking the checkers.

This chapter concludes with Table 9-3. It is a learning aid and a reference to quickly review the main actions required for each step in our problem solving wheel.

Table 9-3. Summary Highlights of the Six-Step Problem Solving Model

STEP 1: Define and select the problem.

Tasks:

- Identify a situational problem topic.
- Dig out and examine data and information concering the extent/nature of the sitauation as it exists in the "current state."
- Use this data/info to describe accurately the "current state"; then write a "desired future state" to define your gap.

Output Requirements to Move to Step 2:

A problem statement written as a gap closing between the "current" and "desired state."

STEP 2: Analyze the problem.

Tasks:

- List possible causes.
- Identify and graphically display probable causes.
- Identify, prioritize, and analyze root causes.

Output Requirements to Move to Step 3:

Root causes selected to work on and, if required, a revised problem statement.

Table 9-3. Summary Highlights of the Six-Step Problem Solving Model (*Continued*)

STEP 3: Generate potential solutions.

Tasks:

- Review problem statement and root causes.
- Generate list of potential solutions.
- Clarify potential solutions.

Output Requirements to Move to Step 4:

A set of potential solutions agreed to.

STEP 4: Select and plan the solution set.

Tasks:

- Determine selection criteria.
- Select/agree on solutions to implement.
- Develop action plan to implement solutions.
- Include both in-process and final measures in your plan to monitor solution progress.

Output Requirements to Move to Step 5:

Final (optimum) solution set decided on and an implementation plan developed.

STEP 5: Implement the solution set.

Tasks:

- Follow implementation plan.
- Use tracking system established in Step 4 to monitor/evaluate both the progress and effectiveness of the solution.
- Implement contingency plans as required.

Output Requirements to Move to Step 6:

The chosen solution set and action plan to solve the identified problem is implemented and monitored through to completion.

STEP 6: Evaluate solution results.

Tasks:

- Compile/display collected data.
- Compare with Step 1 business objective, desired state, and current state.
- Check for new problems or incomplete results created by the solution.
- Recycle through process to address new problems/causes as needed.

Outputs (Recycle If Needed):

How well did the solution set solve the problem? What did we learn? Results documented?

PROBLEM SOLVING I: KEY LEARNING POINTS AND WHAT I WANT TO DO TO IMPROVE MY SKILLS

My Key Learning Points from Chapter Nine:

What I Want to Do Differently:

READER REFLECTIONS AND APPLICATION ACTIVITIES FOR CHAPTER NINE

Reflections

To help you better understand where you might be undermining your problem solving approach, score yourself from 1 to 10 on these four questions.

1. To what extent do I solve problems on an intuitive, gut-feeling basis instead of a methodical, analytical basis?

   (*Great extent*) 1 2 3 4 5 6 7 8 9 10 (*Little extent*)

2. To what extent do I create weak problem definition statements that ignore the "gap closing" elements of a *current state* paired with a *desired future state*?

   (*Great extent*) 1 2 3 4 5 6 7 8 9 10 (*Little extent*)

3. To what extent do my personal needs, goals, objectives, or desires overwhelm and/or color my problem solving processes so I am manipulative versus collaborative?

   (*Great extent*) 1 2 3 4 5 6 7 8 9 10 (*Little extent*)

4. To what extent do I have preconceived solutions that I try to drive home instead of being an open-minded, collaborative problem solver?

   (*Great extent*) 1 2 3 4 5 6 7 8 9 10 (*Little extent*)

Any question scored as 1–5 should be flagged as an area of concern for behavior change using ideas from this chapter.

Leader Application Activity Using Problem Solving I Chapter with an Individual (30 to 35 minutes)

1. Have teammate read Chapter Nine and come prepared with answers to two questions in advance of meeting with you:
   - What did you feel were the most important learning points from the chapter?
   - Why were these key learning points for you?
2. Open the session by letting the teammate share his or her views on the two prework questions above. Actively listen, understand.
3. Share what you found important in the chapter and why these points were key to you.
4. Probe to discover: What might be typical issues faced within his or her team regarding the facilitation of each step of the wheel? How might these facilitation issues be overcome?
5. Is there a gap between your teammate's current utilization/facilitation of the structured problem solving model and where your teammate would like to be? Is the gap based on a knowledge/skills deficiency, a lack of confidence, or both?
6. Mutually agree on a few actions to help close the identified gap. Set a touch-points schedule to review "What is going well?" and "What could be going better?"
7. Follow up on the touch-points schedule, review progress, refine actions and processes that require further practice, and reinforce actions and processes that are being performed well.

Leader Application Activity Using the Problem Solving I Chapter with Your Team (45 minutes)

1. Have all teammates read Chapter Nine and come prepared with answers to two questions in advance of the meeting:
   - What did you feel were the most important learning points from the chapter?
   - Why were these key learning points for you?
2. Open the session by asking teammates to share their views on the two prework questions. Actively listen, understand.
3. Share what you found important in the chapter and why these points were key to you.
4. Probe the team to discover viewpoints and draw conclusions to these specific questions:
   - In what ways could structured team problem solving be applied to help us build a stronger, more productive unit?
   - What do we need to do to promote structured problem solving in our unit and institutionalize it as "the way we do business around here"?
5. How will we monitor and evaluate the use of structured team problem solving in our unit?

NOTES

1. W. G. Dyer, *Contemporary Issues in Management and Organization Development* (Reading, Massachusetts: Addison-Wesley Publishing, 1983), 33.
2. E. Schein, *Process Consultation, Vol. 1, Its Role in Organization Development* (Reading, MA: Addison-Wesley Publishing, 1988), 62–63.
3. T. Baldwin, W. Bommer, R. Rubin, *Developing Management Skills: What Great Managers Know and Do* (New York: McGraw-Hill/ Irwin, 2008), 110.

CHAPTER TEN

TEAM PROBLEM SOLVING II

Structured Methods for Creating, Displaying, and Analyzing Problem Solving Data

CHAPTER OBJECTIVES

- To present a three-step process for managing divergent and convergent thinking
- To present three techniques for displaying data graphically: time charts, bar charts, and pie charts
- To present five techniques for constructing and using analytical tools for problem solving: flow charts, cause-and-effect diagrams, Pareto analysis charts, force field diagrams, and criteria matrices

INTRODUCTION

One of the Japanese experts on quality and employee involvement compares the need for patience and discipline in learning to use basic analytical tools to that of the bamboo farmer. Once the bamboo seed is planted, the farmer must water it every day.

He must do this for four years before the tree breaks ground! But when it does, it grows 60 feet in 90 days!

That's true of analytical tools; it takes a bit of time to learn and apply them, it takes nurturing, it takes practice, and it takes discipline. But once understood and applied properly, they will rapidly increase your analytical power.

This chapter will take you through the practices necessary to facilitate systematic, collaborative problem solving. As covered in Chapter Nine, the second step of the problem solving process is "analyze the problem." The tools presented in this chapter are the means to concrete problem analysis. While a little study and practice may be required for you to get comfortable using these tools, the time required is not too great. One thing is certain: it will be far less than the four years the bamboo farmer has to wait!

The beauty of the analytical tools we will present is their broad application and ease of use. You don't have to be in the second step of a formal problem solving activity with your teammates to take advantage of them. You can apply these tools quickly and easily to many situations. You can use them alone or in a team to organize and clarify data and information, to make a key point more effectively, to lend credibility to your argument, to stimulate thoughts in others, or bring critical thinking to an issue. You should get comfortable using the tools routinely in a variety of situations to promote communications and understanding. These tools are your bamboo seeds; nurture their use and they will grow to be an integral part of your building team power repertoire.

TOOLS FOR GENERATING, SCRUBBING, AND PRIORITIZING INFORMATION LISTS

The story is told that former President Harry S. Truman became quite frustrated with his cautious economic advisors. He stopped one session with them by yelling that he was tired of their wishy-washy, "two-handed" analysis. "On the one hand this, but on the other hand that." Truman, thoroughly agitated, exclaimed, "Would someone please find me a good one-armed economist!"

This anecdote has direct application to information lists. Once a list of items is generated by a group, its members often waste untold hours going around and around in a "two-armed" debate on the consolidation and prioritization of the items: "on one hand this, but on the other hand that." When all is said and done, plenty has been said but little has been done to scrub the list.

The core of exemplary collaborative leadership is the controlled management of information—both verbal and written. Since many group sessions involve the generation of a list of ideas (divergent thinking), coupled with the consolidation and refinement of the items on that list (convergent thinking), mastering the simple three-step process shown in Figure 10-1 will be of immense value when you're required to perform this crucial facilitation activity. The model will help you become a "one-armed facilitator" of list management by showing you how to eliminate rambling and unproductive, circular two-handed encounters among teammates that lead nowhere.

Figure 10-1. List Management: Facilitating Divergent and Convergent Thinking

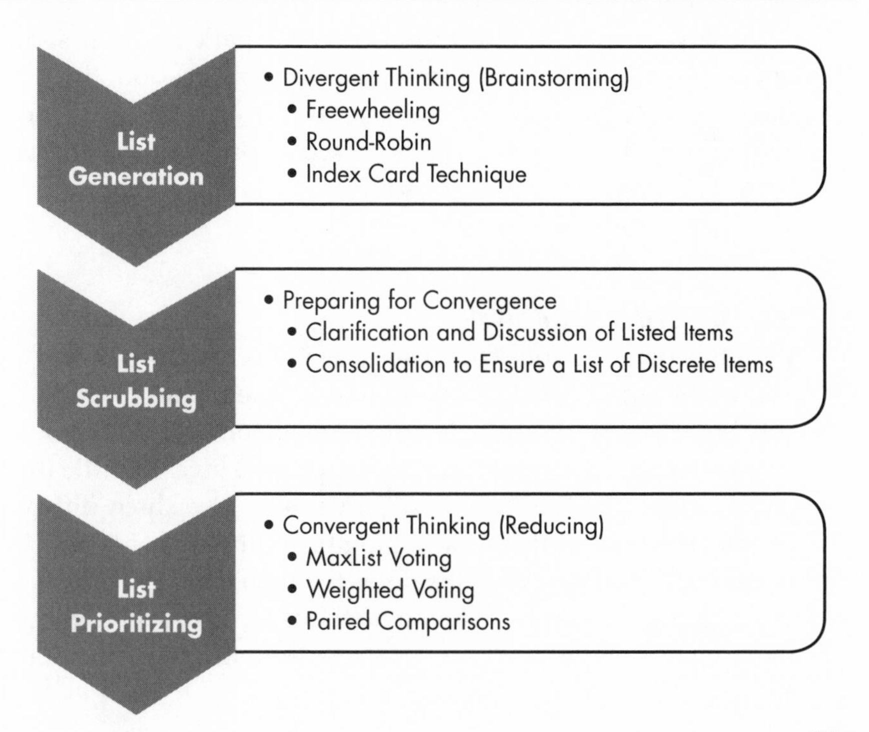

List Generation: Divergence of Ideas through Brainstorming

Freewheeling

This is the most familiar brainstorming approach. Team members call out their ideas spontaneously and the scribe records all of them on a flip chart as quickly as possible and without question. If the group is 10 or more, two scribes may be needed to make certain all ideas are recorded as quickly as possible. Momentum and building on others' ideas are critical during a freewheeling brainstorming session. *Remember to gate-open for the scribe and have that person add ideas along with the rest of the group.*

Round Robin

With this process you, as collaborative leader, go around the table asking each person in turn for an idea. The ideas are recorded on a flip chart as in freewheeling. One cycle around the table is considered one round. A teammate may pass on any round, but this does not preclude that person from contributing an idea on the next or any succeeding round.

Ideas presented in one round tend to spark ideas in subsequent rounds. Before starting the round robin session, the team can agree on the number of rounds it will use, or the activity can continue until all members have passed during a round. *Remember to gate-open for the scribe during each round and have that person add an idea or pass.*

Index Card Technique

With this process, you place a stack of index cards in the middle of the table and invite teammates to print their ideas on them. As collaborative leader and a team member, you will do the same. After a preset time period, collect the printed ideas, shuffle them thoroughly, split the deck into thirds, and give three different people one-third of the deck. Taking turns, these people read the printed ideas while the scribe notes them on a flip chart.

While all three brainstorming methods are excellent divergent thinking techniques, each has its strengths and drawbacks. Table 10-1 highlights these.

Table 10-1. Balance Sheet on Brainstorming Methods

STRENGTHS	DRAWBACKS
Freewheeling:	**Freewheeling:**
➤ Very spontaneous ➤ Tends to be creative ➤ Easy to build on others' ideas ➤ Energizing	➤ Strong-willed teammates may dominate the session ➤ Chaotic; ideas not heard when too many talk at once ➤ Scribe can be overwhelmed trying to note all the ideas coming out at once
Round Robin:	**Round Robin:**
➤ Difficult for one individual to dominate ➤ Process tends to be more focused, less chaotic ➤ Everyone is encouraged to take part ➤ Easy to build on others' ideas	➤ Difficult to wait one's turn ➤ Some loss of energy ➤ Reluctance to pass ➤ Long-winded, rambling ideas presented since person "owns" the floor
Index Card:	**Index Card:**
➤ Anonymity allows sensitive ideas to surface ➤ Provides strong task organization ➤ People have time to think/reflect without interruption ➤ Eliminates both dominance by high-status, aggressive members and pressure to conform to the group	➤ Not possible to build on others' ideas ➤ Some ideas may be poorly written; not understandable ➤ Difficult to clarify ideas ➤ It's a slower process; some people may soon get bored ➤ Easy to be a nonparticipant by turning in blank cards

Key Ground Rules for Brainstorming

1. Stop all initial judgments or premature evaluations of ideas as they are proposed.
2. Be relentless in building a list; do not allow the group to get sidetracked by starting to process the first proposal that seems acceptable or workable.
3. Go for quantity; incomplete or wild ideas are okay.
4. Encourage building on ideas.
5. When brainstorming, don't overlook the scribe or leader acting as facilitator.

List Scrubbing: Preparing for Convergence

Once a raw list of items has been produced using one of the above three methods, these ideas need to be (1) discussed and clarified, and then (2) refined and consolidated to produce a list of distinct items. This process is known as "scrubbing the list."

As the leader facilitating team collaboration, you must ensure that this "scrubbing step" takes place before any action is taken to evaluate and prioritize the list into the vital few. *Scrubbing is always considered the mandatory intermediate step of list management,* because at this stage, your information most likely will be subject to: (1) overlapping (closely related) items, (2) multiple interpretations of some items, and/or (3) misunderstandings of some items. There are two phases to list scrubbing:

Phase 1: Clarification

The purpose of this phase is to enhance understanding and amplification while minimizing influence based on some members' keen verbal skills or higher status levels. It is not the purpose of the discussion/clarification phase to debate the merits of an item, to reach agreement about an item, or to persuade others to adopt your view through repeated advocating. Except for the four ground rules noted next, all other activity is ruled out!

1. Any participant may seek or give clarification about what an item means.
2. Any participant may give reasons he or she is for or against an item.
3. Any participant may ask for more background or an example to gain a better understanding of an item.
4. There will be no obligation to discuss any item, only an assured opportunity for any teammate to initiate a discussion of any item.

Phase 2: Consolidation

In this phase of list scrubbing, acting as a collaborative leader, you must make certain that the final list contains a set of unique, differentiated items so that the final step—reduction and prioritization—is not confounded by duplications, overlaps, or highly related but uncategorized items. As such, you have three collaborative activities to facilitate with the team at this stage.

1. Cross out duplicate items while preserving the one item that best captures the thought (tip: beside the saved item, note all numbers of items crossed out).
2. Set a "bucket heading" that categorizes similar (but distinct) items under a common headline.
3. Reword a statement to encompass several others—beside the newly rewritten statement, note all numbers crossed out and now encompassed by it.

When conducting the consolidation activity, any participant can ask any other group member to give their rationale for a proposed consolidation of two or more items so everyone can better understand the underlying thinking. Once understood, a consolidation proposal may be challenged and a different suggestion offered. However, if two or more participants cannot support a proposed change after a brief discussion, the modification is *not* made. *People cannot be allowed to engage in prolonged argumentation!*

List Prioritization: Convergence of Ideas through Evaluation and Ranking

With a consolidated list of discrete items displayed for everyone to see, you are prepared to move the group into the third, and final, step of list management: evaluation and prioritization. There are three primary processes for reducing a list to the significant few ideas of greatest merit, or, put differently, for promoting convergent thinking. The three processes are MaxList voting, weighted voting, and paired comparisons. The nice thing about these list reduction methods is they can be used independently or in combination.

MaxList Voting

This is a simple way to rank-order the positions and preferences of group members with respect to lengthy "scrubbed" lists. The MaxList (Maximum List) method is particularly useful when your scrubbed list contains roughly 16 or more discrete items. But it can still be used effectively on shorter lists.

With MaxList voting, teammates are asked to make individual judgments about the highest-priority ideas on a master scrubbed list completed in the discussion and consolidation phases earlier. Then these judgments are expressed quantitatively to indicate their relative preferences. The nine points shown here set out the procedural steps for MaxList voting and must be adhered to if this process is to function properly:

1. Before starting the MaxList method, be sure you have neatly rewritten and renumbered flip chart pages containing only the final scrubbed list items.
2. The number of votes given to each participant for allocation among the items on the scrubbed list equals 1½ of the number of items on the list. (See the sidebar on page 225 for an example.)
3. Participants cannot place more than one-third of their total votes on any single idea. (This rule forces all members to vote for at least three options and prevents someone from unduly influencing the proceedings by dumping all votes on a single item.)

4. All assigned votes must be allocated to items on the list.
5. All votes are cast as whole numbers; no fractional votes.
6. Each person writes on a slip of paper only their chosen item numbers with the corresponding votes assigned to those items. (Each person is asked to check their math to make sure all votes are used).
7. The sheets are passed to a single teammate to read out the results.
8. The scribe plainly records all votes on the "scrubbed" flip chart by noting them next to the appropriate items as they are called out.
9. Finally, with the help of the teammates, all the votes are added up. The scribe moves down the list writing the final total for each item and circling it.

For example: If there are 18 items on the list, each participant receives 27 votes to allocate among the 18 items: 1½ x 18 items = 27 votes.

Note 1: If your list is *30 or more* items, you give each participant a maximum of 45 votes regardless of the length of the list.

Note 2: If you have an odd number of items on your final list, just add 1 and carry out the vote calculation as above: 21 items on list, add 1 = 22; 1½ x 22 = 33 votes.

Weighted Voting

This technique is similar to MaxList but is a bit simpler and is particularly useful when your scrubbed list is shorter: containing between 8 and 16 discrete items. The six steps noted here comprise the procedural ground rules for weighted voting, and they must be adhered to if this process is to function properly.

1. Before starting the weighted voting method, be sure you have a neatly rewritten flip chart page that contains only the final scrubbed list items as previously agreed upon.

2. Each participant is given just three weighted votes: a 5-point vote, a 3-point vote, and a one-point vote to allocate among the items on the scrubbed list.
3. All three weighted votes must be allocated to items on the list.
4. All votes must be cast as 5, 3, and 1; there are no other options.
5. Team members write on a slip of paper only their three item numbers with corresponding votes assigned to them.
6. The sheets are passed to a single person to read out the results.
7. The scribe plainly records all votes next to the items on the flip chart and, with the help of the team, tallies the vote total for each item and circles each item's total.

For example: Team members place their 5-point vote on their most preferred item, their 3-point vote on their second most preferred item, and their 1-point vote on their third most preferred item.

Paired Comparisons

The paired comparisons method will help your team quantify everyone's preferences without any preassigned number of votes given to anyone. Each option on the list goes head-to-head against every other option in a sequence of "paired comparisons." With each paired face-off, teammates are asked to vote for their most preferred option in that pair. Votes are recorded and totaled after all possible comparisons have been made. (See the sidebar example on page 227.)

Using paired comparisons is an excellent technique when the scrubbed list is small—containing three to eight discrete items. Because the number of comparisons accelerates rapidly with each additional item put on the list, it is best to reserve paired comparisons for short lists by following these six procedural steps.

1. Before starting the paired comparisons method, be sure you have a neatly rewritten flip chart page that contains only the final list of scrubbed items previously agreed upon.

2. With paired comparisons, each item on the master scrubbed list is paired up, in turn, with every other item on the list, and voted on. See the sidebar following this numbered list.

For example: Take a list of four colors to be evaluated by the team in order to choose its favorite color.

Red → Round 1 *(Three comparisons):* Red is compared to blue; red is compared to green; red is compared to yellow.
Blue → Round 2 *(Two comparisons):* Blue is compared to green; blue is compared to yellow.
Green → Round 3 *(One comparison):* Green is compared to yellow.
Yellow.

Assuming a total of eight team members including yourself, you would lead the paired comparisons voting process saying something like this:

"For our first comparison, our choice of colors is limited to red and blue. Given just that choice, how many of you would prefer red? Raise your hand if you prefer red. Okay, four hands are up including my own, so would the scribe write a 4 next to red and a 4 next to blue?" (By not voting for red in the paired comparison, those people have voted for blue.)

"For our second comparison, the choice now is between red and green. How many people prefer red to green? All right, I count six hands including my own, so would the scribe write a 6 next to red and a 2 next to green?"

"For our last comparison in round one, the choice is limited to just red and yellow. Given that choice, how many of you would prefer red? Raise your hand if you prefer red. Okay, five hands are up, so would the scribe write a 5 next to red and a 3 next to yellow? (And so it would go, following the same process for the two comparisons in round two and the final comparison in round three.)

3. As you can see, with each comparison, everyone must make a choice between just those two items and vote for one alternative or the other.
4. When finished, every item will have been compared to every other alternative in a "paired face-off," ensuring a thorough dispensation of each item.
5. Votes are tallied by the scribe next to each item as each comparison is made.
6. Totals are then computed for each item once the paired comparison process has been completed and the total votes for each option is circled.

TOOLS FOR DISPLAYING DATA GRAPHICALLY

It is impossible to pick up a newspaper, news magazine, or business magazine without seeing a graphic display of some sort to make it easier for the reader to grasp a piece of information. A picture certainly is worth a thousand words, and that is what makes the three graphic displays shown here so powerful.

For example: Take a list of four colors to be evaluated by the team in order to choose its favorite color.

1. Red
2. Blue
3. Green
4. Yellow

Assuming there is a total of eight team members including yourself, you would lead the paired comparisons voting process, saying something like this:

"For our first comparison, our choice of colors is limited to red and blue. Given just that choice, how many of you would prefer red? Raise your hand if you prefer red. Okay, four hands are up including my own so, would the scribe write a 4 next to red and a 4 next to blue." (By not voting for red in the paired comparison, those people have voted for blue.)

"For our second comparison, the choice now is between red and green. How many people prefer red to green? All right, I count six hands including my own, so would the scribe write a 6 next to red and a 2 next to green."

"For our last comparison in round one, the choice is limited to just red and yellow. Given that choice, how many of you would prefer red? Raise your hand if you prefer red. Okay, five hands are up, so would the scribe write a 5 next to red and a 3 next to yellow." (And so it would go, following the same process for the two comparisons in round two, and the final comparison in round three.)

Use them to help you make points and to visually reinforce your messages.

If you or a subgroup have collected data through interviews, surveys, observations, benchmarking, library research, or some other means, displaying the results to others is particularly important. During analysis and decision making, trends and sequences are often more evident, and comparisons more easily made, from graphic representations of data. Three common tools used for displaying data graphically are *time charts, pie charts,* and *bar charts*. These graphics are among the simplest and best formats for communicating data. And with today's software, each can be created quickly and simply by plugging your raw data into an Excel spreadsheet or other similar program.

Time Charts

Use a time chart to summarize and display changes over time. The bottom, or horizontal, axis shows the time intervals; the left side, or vertical, axis presents your number scale indicating frequency. The plotted points are connected by a solid line. If the same time period is used, multiple data sets can be displayed on one chart, as Figure 10-2 shows.

Figure 10-2. Time Chart Example

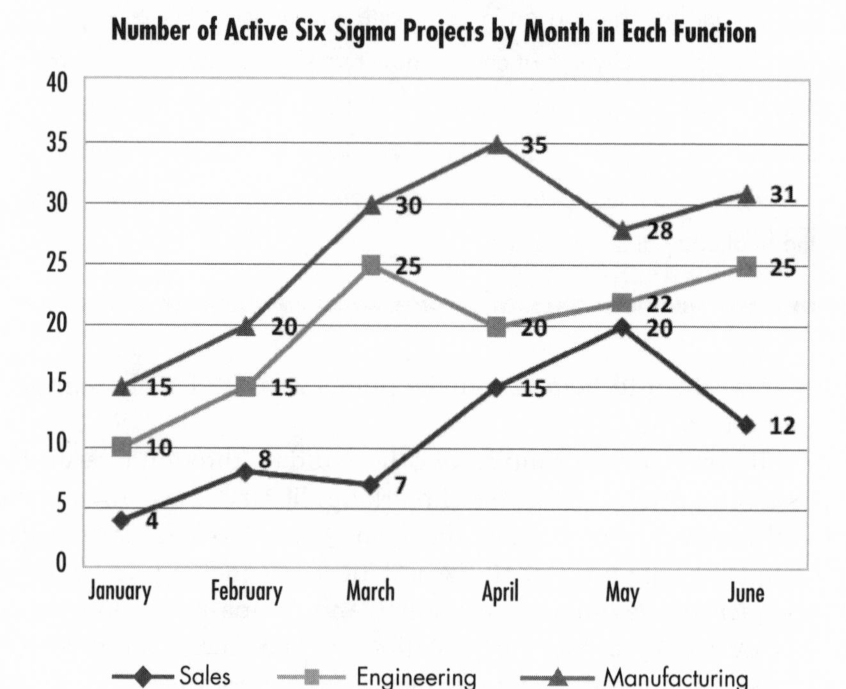

Pie Charts

Use a pie chart to show the relationship of parts of a whole to each other, and of each part to the whole. Show comparisons among quantities by dividing a circle into "pie" wedges.

The whole "pie" equals 100 percent, so the parts of the pie must add up to 100 percent. The size of each wedge should be proportional to its percentage of the whole. Pie charts are easily interpreted and can present data effectively and efficiently (Figure 10-3).

Figure 10-3. Pie Chart Example

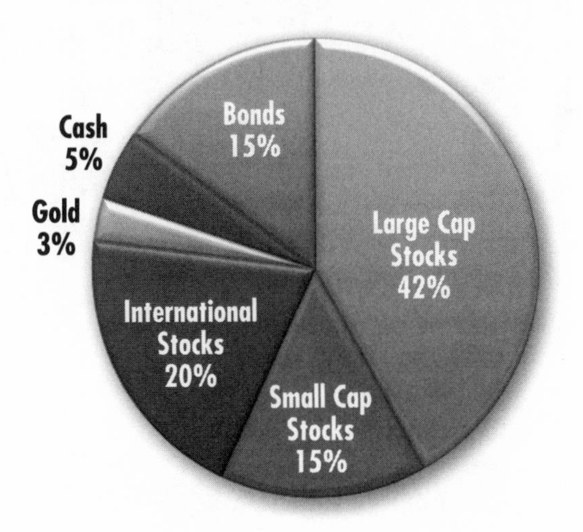

Bar Charts

Use bar charts to show a comparison of quantities of like items (lost customers by region, defects by shift, percent of students in a district by grade level, and so on). The different quantities are indicated by the length of the parallel bars used to represent them. Bars may run vertically or horizontally. For a vertical bar chart, use the horizontal axis to show the different regions, shifts, grade levels, or whatever factor you are working with. Use the vertical axis for your number scale. To create a horizontal bar chart, reverse the axes (Figure 10-4).

Figure 10-4. Bar Chart Example

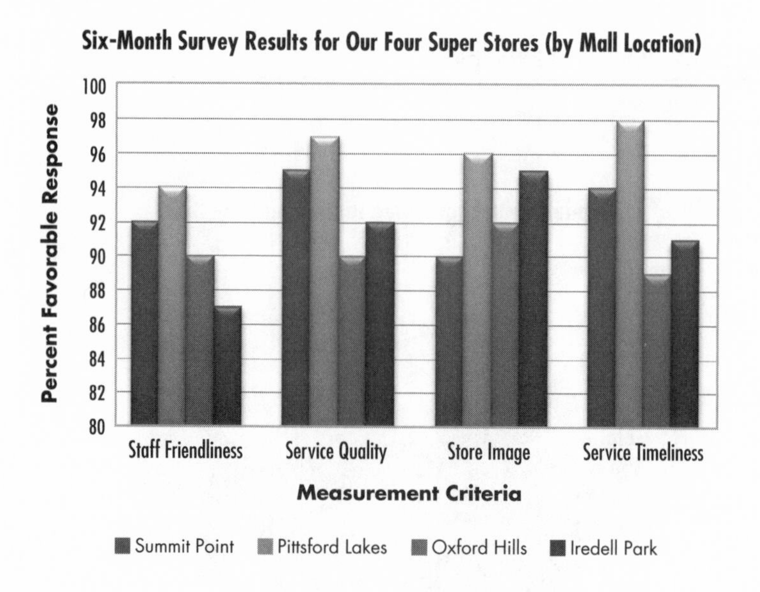

TOOLS FOR ANALYZING DATA GRAPHICALLY

The tools described next are used to analyze data, as well as to display it. That is, the creation of any of these graphic charts or diagrams is itself an analytic process.

The Flow Chart

The value of the flow chart in problem solving cannot be overstated. A well-constructed flow chart can clear up roughly 30 percent of your problem solving issues—especially work process issues—without any further activity. A flow chart is not complicated. It is simply a pictorial representation of the steps leading to an output—it documents a work process.

When doing a flow chart, you must depict the function as it operates in reality, not how you believe your boss thinks it operates or how the team would like it to operate. *Work process reality* is best achieved by making certain that one or two people intimately involved in the process are members of the flow charting group. Don't bring in the person who designed the original process. Don't bring in the theoretical expert on this process. Don't get someone who knows how the process should run. It is important that you find and utilize people who actually live with the work process because they are the only ones who know how the process really functions. For a complex process, people drawn from several functions may be needed to create the correct flow.

The value of flow charting is that it identifies improvement possibilities, unnecessary steps, unclear decision points, not enough decision points, too many movement steps, bottlenecks, pressure points, and the like. Creating a flow chart requires you to lead your teammates through the following seven steps.

1. Agree on the work process needing analysis.
2. Agree on the start and end points of the process; this helps keep the team focused.
3. Brainstorm and write the major steps in the process on Post-it notes; do not worry about sequence at this point.
4. Put the Post-it notes in a rough sequence; keep moving them around, adding and taking away steps as needed to get the real process.
5. When done, draw the appropriate symbols around the Post-it(s) to define the steps.
6. Connect steps with arrows to show the flow.
7. Test the chart to be sure all the key steps have been included in proper sequence.

One final tip for sharpening your analysis is to draw two flow charts. After you develop one depicting reality, draw a second illustrating how the process should operate. Then by comparing and contrasting the current reality with the desired goal, pinch points and problem areas in the process are often easier to isolate and understand.

Figure 10-5 shows a completed flow chart for a departmental expense reporting process. As the legend inset indicates, there are four universal symbols—ovals, rectangles, diamonds, and arrows—that are used to depict what is taking place.

Figure 10-5. Flow Chart Example

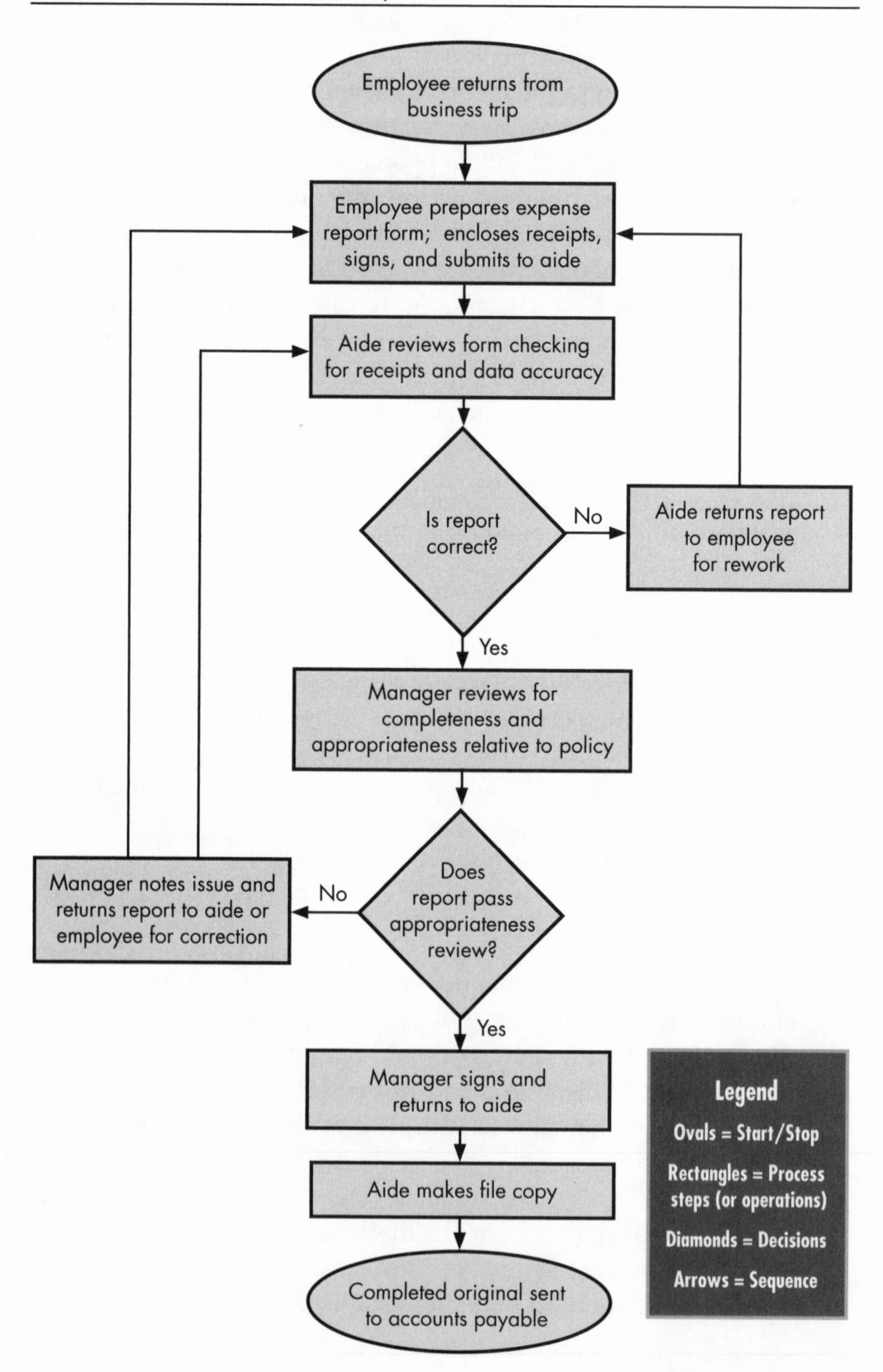

The Cause-and-Effect Diagram

The cause-and-effect diagram is also referred to as either the Ishikawa diagram (because it is based on a method developed by Dr. Kaoru Ishikawa—teacher, expert, and author of several books in the field of quality control), or the fishbone diagram (because, when developed, it looks like a fish skeleton).

Cause-and-effect analysis is a powerful technique for triggering ideas, recording brainstormed ideas, and systematically examining an effect and the various causes that create or contribute to the effect. Facilitating the creation of a cause-and-effect diagram is fairly straightforward. Remember, for every effect there is likely to be several major causes and a number of subcauses.

Let's take an example and trace through the construction of a cause-and-effect diagram.

The executive vice president of manufacturing operations in a high-technology enterprise was alarmed to discover that actual manufacturing costs were way out of line compared to its planned standard costs. So he handpicked a senior level, cross-functional team to address the situation.

One of the analytical tools this team used was a cause-and-effect diagram. The leader initiated it by drawing a large box at the far right side of the flip chart. The team worked hard to get a quantifiable effect written in this box. This represented the *head of the fish*. A *spine* was drawn, and *four major bones of the fish* were drawn off the spine, indicating primary influences or causes of the effect. The teammates spent time discussing and then agreeing on four causal labels for each of the four main bones. Then the team worked together and reached consensus on the diagram shown in Figure 10-6.

Since the building of an Ishikawa diagram always begins with and is driven by the defined effect statement, get consensus on it before moving forward. In many instances this may be a no-brainer. However, don't assume everyone sees the effect the same way. Ask for a proposal, or in your role as collaborative leader suggest one yourself. Note it on a flip chart and test for consensus to make sure everyone is on board. If not, work with your teammates to modify the proposed statement until a con-

Figure 10-6. Cause-and-Effect Diagram Example

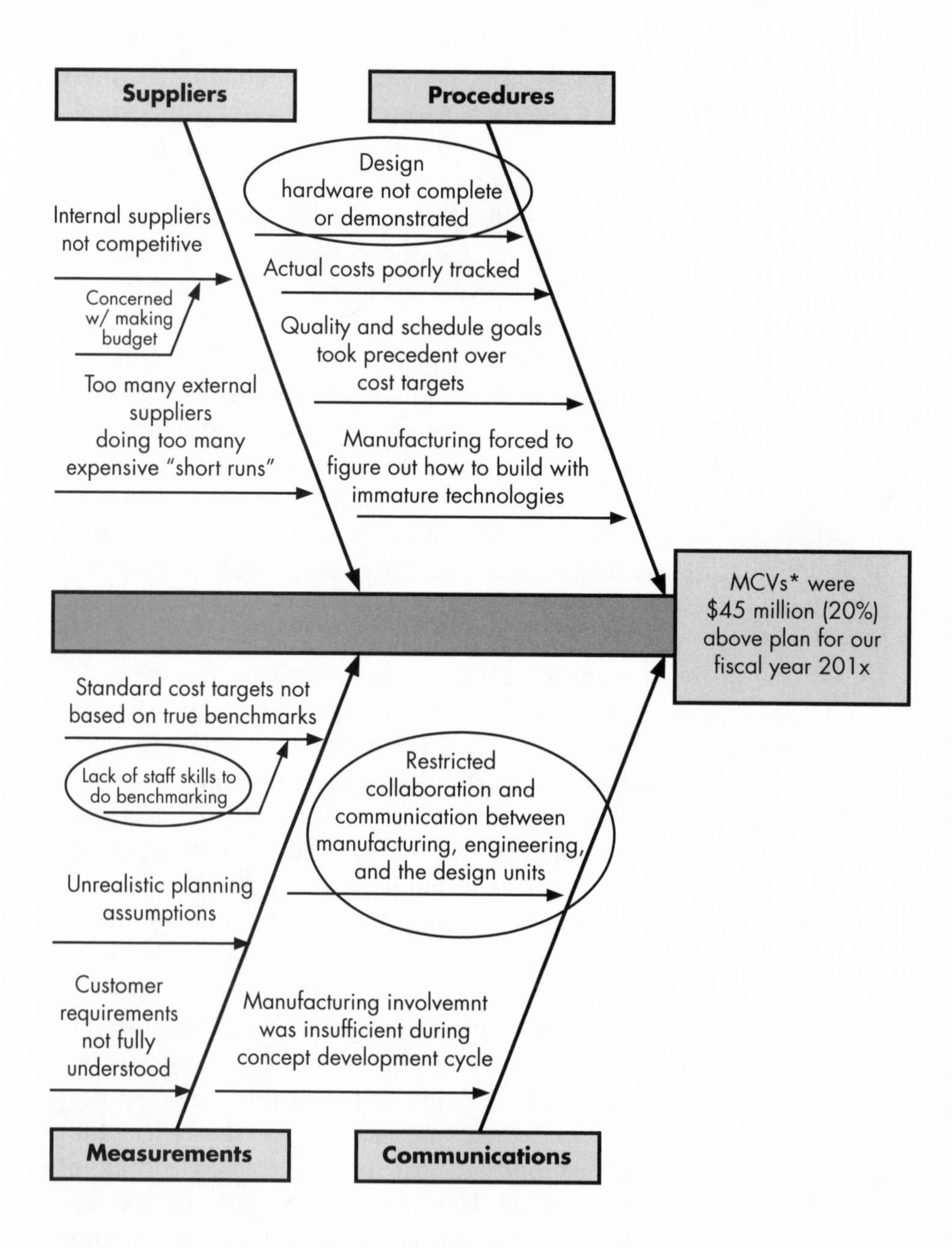

* MCVs = manufacturing cost variances. This is the difference between the actual cost incurred and the average cost planned to be incurred (standard cost) for a given part, subassembly, or end product.

sensus is reached. My experience over many situations is that this is usually a routine, noncontroversial step.

Often it's difficult to work out a way of classifying the primary causes—the boxes labeling the main bones of your fishbone. Table 10-2 highlights three sets of common primary causes that will serve your classification needs many times when initiating and labeling the main bones.

Table 10-2. Three Sets of Main Causes

4Ms	**4Ps**	**GRRP**
Manpower	Policies	Goals
Machines	Procedures	Roles
Methods	People	Relationships
Materials	Plant	Procedures

The 4Ms are the most widely used, the 4Ps are helpful when examining administrative areas, and GRRP is a valuable classification set when looking at the internal functioning or operations of a team of people. There is one other universal cause, *environment,* that is helpful and can be used with any of the major groupings.

While these three classification systems are universal, any major causes that emerge or help you think creatively—like those in Figure 10-6—is value added to this process. Encourage people to suggest any primary cause they feel strongly about. However, keep the primary bones to four to six items so the whole process does not get out of control. Don't be afraid to mix items from the three classifications if that helps everyone to better visualize and understand the cause-and-effect relationships. Nothing is carved in stone.

Creating a cause-and-effect diagram requires you to lead your teammates through the following nine simple steps.

1. Prepare to develop an easy-to-see diagram by taping a large sheet of paper horizontally on the wall, or use a white board.
2. Place the problem statement (quantified if possible) in a box on the right. It must represent some aspect of the current state.
3. Define the primary categories by choosing one of the universal sets, developing your own set, or mixing some together.

4. Use the primary cause categories to brainstorm possible causes. For example, "What is it in procedures that is causing . . . ?"
5. Write each brainstormed cause in a few words on a Post-it note. Stick each note next to the appropriate major category rib. For example, next to the major rib for "measurements," put a note: "Standard cost targets not based on true benchmarks."
6. For each cause ask, "Why does this happen?" For example, "Why weren't standard cost targets based on benchmarks?" Answer: "Because we lack staff skills to do benchmarking." Place this answer on a self-stick note and show it as a subcause branch off of "Standard cost targets not based on true benchmarks."
7. Keep digging for other root causes. A potential root cause is any cause without subcauses or the final subcause for an item.
8. Examine and discuss the other causes; single out other likely root causes by drawing a circle around each one of them.
9. With the significant few identified (circled), test consensus to make certain everyone supports the choices.

Any of the initial root causes uncovered and selected through your cause-and-effect diagram must now be validated by collecting supporting data. If the data does not substantiate the soundness of one or more root causes selected, you and the team will have to revisit your diagram, choose other less obvious causes, and then collect data on these.

Finally, the cause-and-effect diagram can be linked directly to a flow chart. Once you have identified a bottleneck, an unclear decision point, pressure point, or the like with the flow chart, you can write the identified bottleneck as an effect statement at the head of the fish. Then you can work through a cause-and-effect analysis as described, and unearth the root causes for your bottleneck.

Pareto Analysis

A Pareto analysis, displayed as a Pareto chart, rank-orders data and shows which factors in a situation occur most frequently. This ranking is useful in determining aspects of a problem or situation that will yield the greatest payback for the time, money, or effort invested. The data is displayed on a chart resembling a bar chart but arrayed in descending order. Pareto analysis is a technique that separates the vital few from the trivial many.

Vilfredo Pareto, a nineteenth century Italian economist, wanted to show that the majority of wealth was held by very few people—that wealth was unequally distributed. The familiar 80-20 rule (80 percent of the wealth is held by 20 percent of the people) is an example of Pareto analysis. Such analyses are often used to rank-order quality problems and put them in perspective according to their magnitude or prevalence. For each problem type, the number of occurrences is recorded on a check sheet for a sample of items over a given time period. The results, depicted in a bar graph (Figure 10-7), are ranked in descending order of incidence.

By its steepness, the curve of cumulative percentages always visually emphasizes the importance of each item. Pareto diagrams can aid decision making by demonstrating where to begin an improvement activity or how to allocate limited resources to an improvement initiative. Figure 10-7 shows that 70 percent of the paint defects result from three primary causes. Concentrating resources to fix those three areas and then creating other Pareto charts with new data over time will prove whether the implemented solutions reduced the relative frequency of paint defects and/or their costs.

Figure 10-7. Pareto Analysis Example

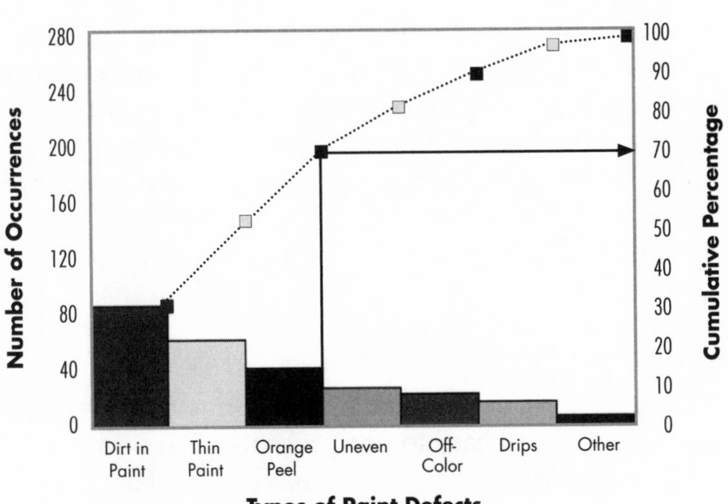

Coupling a *dollar cost per defect* to your Pareto analysis adds even more horsepower. Let's say, for example, that of the three primary problems, dirt in paint costs you $250 per unit to redo, orange peel costs $175 per item to fix, and thin paint costs $50 per item to repair. In attacking the problems, it's clear that eliminating the "dirt in paint" problem is your number one priority. However, fixing the "orange peel problem" would be your number two priority due to its high defect repair cost. Your third priority would be to tackle the "thin paint problem" because—while higher in frequency versus orange peel—it costs much less per item to fix that particular defect.

The process for constructing a Pareto diagram involves 10 straightforward steps, as follows:

1. Use a checklist to collect the frequency of events occurring across a set of categories for some period of time.
2. Arrange the raw data in order from largest frequency of occurrence to smallest.
3. Calculate the total number of occurrences.
4. Compute the cumulative percent.
5. Draw two vertical axes.
6. Scale the left-hand vertical axis for frequency (zero to the total number of occurrences).
7. Scale the right-hand vertical axis for cumulative percent (zero to 100 percent).
8. For items 6 and 7, make sure that the two axes are drawn to a common scale (e.g., 100 percent is opposite the total frequency, 50 percent is opposite the halfway point, and so forth).
9. Working from left to right, construct a bar for each category, with height indicating frequency. Work in descending order, starting with largest category.
10. Plot the cumulative percent line as shown in our example chart.

Force Field Analysis

This technique, introduced into management theory by Kurt Lewin, is logical and easy to comprehend, even for people who have never constructed a force field before. Another virtue of the force field is its applicability to a wide variety of problem solving situations. A completed force field for a sales improvement problem is represented in Figure 10-8.

Figure 10-8. Force Field Analysis Example

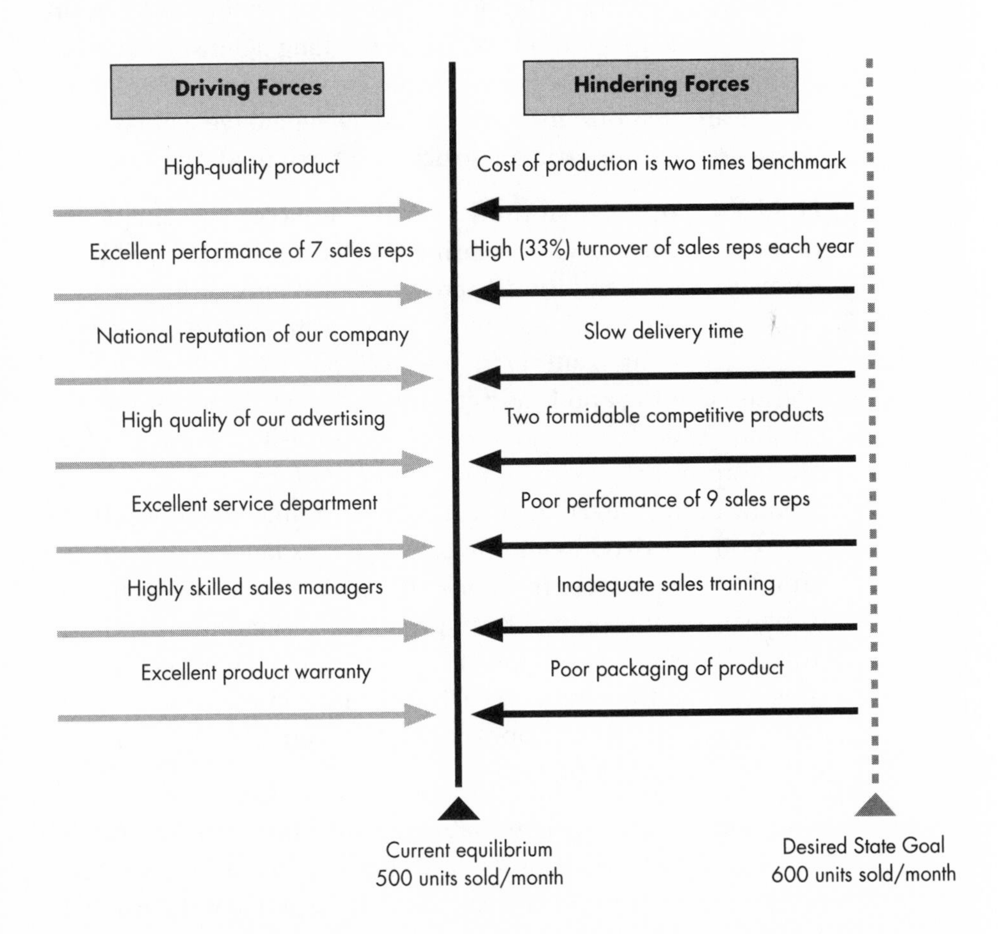

Lewin looked upon a level of performance within an organizational setting—production, sales, customer satisfaction, defects, trust, collaboration, and so on—not as a static habit or custom but rather as a dynamic balance of forces working in opposite directions. Driving forces were seen as facilitating and powering the situation to change to a higher equilibrium point. Hindering forces were seen as countering the driving forces by restricting and restraining movement to a higher equilibrium point. Therefore, the present state of affairs for a given situation is in equilibrium because it is maintained by a variety of offsetting forces that "keep things the way they are" or "keep us behaving in our customary ways."

Regarding problem solving, force field analysis holds that a problem arises out of the stationary product of driving and hindering forces. Thus the two sets of forces working against each other cause a problem because there is a gap between our current state, where we are, and our desired future state, where we want to be.

There are five steps for constructing a force field diagram.

1. Hold a discussion that produces two very concise statements (five to 10 words each) declaring the "current state equilibrium situation" and the "desired state goal situation."
2. Draw the framework and write the two statements in place as shown in our example in Figure 10-8.
3. Identify driving and hindering forces via a brainstorming session. As each is given, write it on a Post-it note and stick it on the appropriate side of the *current equilibrium* line.
4. Review both sides of the chart; discuss, refine, and consolidate both the driving and hindering forces list. Don't be afraid to move forces from one side of the chart to the other as new insight is gained throughout the discussion.
5. You are finished when you and the team have reached a consensus on all the forces on each side of the current equilibrium line (e.g., all the opposing forces deemed to be creating the present situation).

With the force field complete, the question now becomes: Which forces do we focus on for change? According to force field analysis, it is important to look at both sides of the issue. If you tackle problems only by trying to decrease some hindering forces, or by trying to increase some driving forces, you are less likely to improve performance than if you analyze both sides of the situation simultaneously.

In our example, if you only tried to decrease the hindering effect of "nine poor performing sales reps" by getting rid of them, you would run the risk of alienating some of the better performers and might end up still selling only 500 units a month. Or if you increase one of the driving forces, you might well run into increased resistance from one of the hindering forces. The result of such one-sided analysis could be increased tension in the system, but no upward movement in performance.

As you examine your force field, look for combinations from both sides as sets of linked forces to be altered simultaneously. For example, in our illustration you might see the opportunity to take the skills of selected sales managers and use them more creatively by having these people work more intensely with several of the poorly performing sales reps to increase their skills. Another combination example, for linking two opposite forces, might be to improve the quality of sales training (a hindering force) by using the skills of your seven best sales reps (a driving force) to help design a sales training workshop and act as trainers for the rest of the staff.

Questions to help you maximize your analysis as you proceed to examine both sides of your chart:

- Can we turn this hindering force into a driving force?
- Which hindering forces would be the hardest (or easiest) to eliminate or reduce and why?
- How much control do we have over this driving force?
- Could this driving force be applied in other ways?
- Are there other helping forces not listed?

If you can't work opposing forces in tandem, it is best to work on eliminating or reducing hindering forces rather than trying to overpower them by increasing the strength of driving forces. The latter approach increases pressure in the system. It is like trying to drive your car by pressing down harder on the accelerator (increasing a driving force) while keeping your other foot firmly on the brake (increasing resistance from a hindering force): lots of noise, heat, and smoke, but no movement. But, as you ease your foot off the break, without pushing any harder on the accelerator, the car begins to move forward.

Criteria Matrix

The criteria matrix is a valuable tool for evaluating decision alternatives against a set of defined evaluation criteria. By evaluating alternatives based on how each one stacks up against the set of predetermined criteria, a numerical value for each alternative can be identified. These values can then be compared to create a priority ranking among the alternatives. The tool is important because it treats the criteria independently, helping to avoid the common mistake of allowing one or two individual criteria to be batched together and supersede all the rest.

The criteria matrix itself is constructed with a set of review criteria—with each having its own rating scale—listed across the top of the matrix, and the various alternatives to be rated noted down the left side.

The matrix shown in Figure 10-9 is particularly useful in Step 1 of our problem solving process when you and your teammates have identified and defined a number of important problems to solve. However, all team members know the team does not have the process capability to take them on all at once, so you lead the participants in developing a matrix, using this set of criteria shown in Figure 10-9, to sort through the problems and select the one or two highest gain problems to attack first.

A criteria matrix also works nicely in Step 4 of our problem solving process when you and the team are trying to select the most practical solutions to implement without compromising your ability to "do the job right." Without using a criteria matrix, teams often become overly ambitious and attempt to implement too many solutions at once, or try to implement a complex solution beyond its capability to handle properly. In either case, the result invariably is failure, which sours the participants on team problem solving because many weeks or months of hard work moving through the first three steps are wasted. To combat this, the solution prioritization matrix shown in Figure 10-10 is invaluable. While some criteria stay the same as before, others are replaced or redefined because solution prioritization requirements differ from problem selection.

Figure 10-9. Problem Prioritization: Criteria Matrix Example

Evaluation Criteria / Options	CONTROL Little Great 1----------10	IMPORTANCE Little Great 1----------10	DIFFICULTY Great Little 1----------10	TIME Great Little 1----------10	RETURN ON INVESTMENT Low High 1----------10	RESOURCE REQUIREM'T High Low 1----------10
Alternative 1						
Alternative 2						
Alternative 3						
Etc.						
Etc.						
Etc.						

Control: To what extent does this team have the authority and/or influence to reduce or close the problem gap?

Importance: How much does it matter whether this problem is solved?

Difficulty: What is the degree of team effort required to work this problem through to solution?

Time: How long will it take to resolve this problem?

ROI: What is the magnitude of the expected payoff from solving the problem?

Resources: What amount of resources (people, money, equipment, etc.) will be required to solve the problem?

Figure 10-10. Solution Prioritization: Criteria Matrix Example

Evaluation Criteria / Options	CONTROL Little 1----------10 Great	APPROPRIATENESS Low 1----------10 High	ACCEPTABILITY Little 1----------10 Great	TIME Great 1----------10 Little	RETURN ON INVESTMENT Low 1----------10 High	RESOURCE ACCESS Little 1----------10 Great
Alternative 1						
Alternative 2						
Alternative 3						
Etc.						
Etc.						
Etc.						

Control: To what extent can implementation of this solution be managed or extensively influenced by the team?

Appropriateness: To what degree does this solution close the gap between the current and the desired state?

Acceptability: To what extent will the people impacted buy into and commit to the solution?

Time: How long will it take to implement this solution?

ROI: What is the magnitude of the expected payoff from implementing this solution?

Resource Access: To what extent are the key resources (people, money, equipment, etc.) available for implementation?

The construction and use of any criteria matrix is a straightforward process comprised of five steps, as follows:

1. Lead team members in a general discussion about each of the different alternatives. Have them chat about the evaluation criteria in relation to the alternatives without trying to make any formal evaluations. Individual perceptions and views are shared openly to make certain everyone understands the alternatives and the criteria in the same way.
2. Next, ask team members to note individually on a slip of paper their "1-to-10" evaluation for each alternative against each of the six criteria listed across the top. This is done silently.
3. The individual numerical evaluations are shared. These are written in the appropriate cells of the matrix, which have been predrawn on a flip chart.
4. If two or more alternatives are tied for first place—or are within 15 points of each other—more discussion around just those alternatives would have to take place. Consensus should be used as the means for setting the proper priority.
5. An important point: This is not an either/or situation. Several alternatives can be selected and combined into a more comprehensive or longer term solution. You can combine and rearrange a number of individual solutions using a summary of each alternative's score as an indication of the priority of different sets of solutions.

A completed solution prioritization matrix is shown in Figure 10-11. Note that the second and fifth alternatives are within 15 points of each other; therefore, key point 4 comes into play. Even when numerically rating alternatives in a criteria matrix, sometimes final results are too close to differentiate them. Extended debate and discussion are required to sort out things and make a final decision. The numbers cannot rule alone!

After debate and discussion, the team in our example might end up deciding to: (1) select item 2 over 5, (2) select item 5 over 2, or (3) combine items 2 and 5 into a new solution that draws from both. The main point is that picking alternative 5 over 2

Figure 10-11. A Completed Solution Prioritization Criteria Matrix by a Seven-Person Team

Evaluation Criteria / Options	CONTROL Little 1----------10 Great	APPROPRIATENESS Low 1----------10 High	ACCEPTABILITY Little 1----------10 Great	TIME High 1----------10 Low	RETURN ON INVESTMENT Low 1----------10 High	RESOURCE ACCESS Little 1----------10 Great
Alternative 1 – 244 pts.	4,4,5,8,5,3,5	6,6,4,8,8,7,6	8,8,8,8,9,7,8	3,4,5,6,6,3,2	9,9,8,8,8,8,8	4,4,3,2,2,2,5
Alternative 2 – 284 pts.	8,7,7,8,9,9,7	7,6,6,6,6,6,9	7,8,7,8,9,6,6	3,4,4,3,3,2,2	8,9,9,9,9,6,7	6,6,7,8,9,9,9
Alternative 3 – 216 pts.	3,4,5,5,4,6,3	8,8,7,8,8,7,8	7,6,5,4,3,5,4	4,3,3,2,2,4,5	9,9,9,8,9,7,8	1,2,2,2,3,3,3
Alternative 4 – 133 pts.	3,2,2,1,4,5,2	2,2,1,1,3,4,3	6,4,4,3,6,4,3	5,2,3,3,2,1,2	4,4,4,3,3,5,2	5,4,4,3,3,3,3
Alternative 5 – 295 pts.	7,8,7,8,9,6,6	8,9,9,9,9,6,7	7,6,6,7,8,8,6	6,6,7,8,9,9,9	7,6,7,4,5,5,5	7,6,6,6,6,6,9
Alternative 6 – 156 pts.	4,2,2,1,4,5,3	3,3,2,1,4,4,3	7,5,4,4,7,5,3	5,2,3,4,3,1,4	4,4,4,5,3,7,2	5,4,4,6,3,7,3

simply because it is 11 points higher on the criteria matrix would be wrong. Extended dialogue and a meeting of the minds is mandatory when "it is too close to call" numerically.

In essence, the criteria matrix is an excellent tool for evaluating any set of workplace strategies. Instead of not having enough alternative strategic approaches, teams typically find they have too many. The criteria matrix can help you do a quantitative analysis and cull the list to a single choice, or at least narrow it down to the vital few for further deliberation.

The total basket of 12 defined criteria used in our "problem selection" and our "solution selection" matrices likely will be all you'll ever need. These criteria are discriminating and hit the most critical considerations for most decision-making situations. Still, like many tools in this book, there is flexibility. Feel free to mix and match criteria from the pool of 12 to meet your needs in whatever decision situation you face. If you need to create a criterion not in our pool of 12, do so. Just make sure it is precisely defined and everyone understands it the same way. Finally, there is no magic in using six criteria in your matrix. If three, four, or seven works best for your situation, use that number.

A CLOSING POINT

The professor had just handed out the final exam papers to her students when her teaching assistant rushed up to the front of the room in panic.

"Professor Sharkey, you've made a terrible mistake. This is the same test you gave last year; everyone will know the answers."

"Not to worry," replied the professor. "The answers are different this year!"

So it is with our structured problem solving model. The steps in the process, their order, the questions asked, and the tools used may all be the same, but the answers—the solutions derived—will vary greatly from application to application. That is the power of the problem solving tools and processes now at your disposal.

PROBLEM SOLVING II: KEY LEARNING POINTS AND WHAT I WANT TO DO DIFFERENTLY TO IMPROVE

My Key Learning Points from Chapter Ten:

What I Want to Do Differently:

READER REFLECTIONS AND APPLICATION ACTIVITIES FOR CHAPTER TEN

Reflections

By understanding and having used the "eight fundamental tools" as described in Chapter Ten, you will be in an excellent position to question and inspect the application of these same tools by others in teams you are managing but are not directly part of. There are five incisive questions that will reveal the extent to which a team's thinking—in analyzing their problem and evaluating their alternatives—adhered to a systematic process. You need to reflect on and ask the following questions when reviewing team projects:

1. What problem triggered this project and recommendations? Was it clearly a high-priority deviation?
2. What were the probable causes of this problem? Was causation identified? How was it verified? Were other causes looked at?
3. How was the decision made? What criteria or decision factors were used? Where did they come from? Were they specified in some way? Were they prioritized; if so, on what basis? How were all these judgments combined to reach a decision?

4. What weaknesses do you see in the alternative you are recommending?
5. At what point in your analysis did you come to regard your recommended alternative as the best one?

Leader Application Activity: Problem Solving II with an Individual (30 to 35 minutes)

1. Have teammate read Chapter Ten and come prepared with answers to two questions in advance of meeting with you:
   - What did you feel were the most important learning points in Problem Solving II?
   - Why were these key learning points for you?
2. Open session by letting the teammate share his or her views on the two prework questions above. Actively listen and understand.
3. Share what you found important in the chapter and why these points were key for you.
4. Probe to discover: What is a key business problem (defined as a gap) being faced by the teammate? Which tools might be used to analyze the issue and dig out root causes? Why are these tools the most appropriate ones in this case?
5. Mutually select the most appropriate tool(s) for the problem situation and discuss its proper application to the problem.
6. Agree on actions to help the teammate close the identified gap using the chosen tool(s). Set a touch points schedule.
7. Follow up with touch points and review progress using the "Reflection" questions. Work together to improve tool application.

Leader Application Activity: Problem Solving II with Your Team (45 minutes)

1. Have all teammates read Chapter Ten and come prepared with answers to two questions in advance of the meeting:
   - What did you feel were the most important learning points in the chapter?
   - Why were these key learning points for you?

2. Open the session by asking teammates to share their views on the two prework questions. Actively listen, understand.
3. Share what you found important in the chapter and why these points were key for you.
4. Probe the team to discover viewpoints and draw conclusions to these three questions:

   - In what ways could these tools be applied to help reduce cost and waste in our unit?
   - How could we use these tools to remove longstanding customer complaints where our unit's outputs are part of the reason for their complaints?
   - In what ways could we use these tools to develop new products or services, or to improve current ones, that our unit provides?

PART III

A Collaborative Leader in Action Building a Collaborative Partnership

CHAPTER ELEVEN

THE BUDGET CUT

A Case Study Integrating What Has Been Learned

CHAPTER OBJECTIVES

- To use a workplace setting to show collaborative leadership in action
- To demonstrate the application of collaborative leadership tools, techniques, and processes presented throughout this book to build a collaborative partnership
- To highlight the power of collaboration in the face of serious conflict

INTRODUCTION

The case you are about to read is the transcript of a meeting involving Mel and his six direct reports. Mel is president of the restaurant kitchen products division, one of six divisions comprising a company that has a well-established name in its marketplace for designing, manufacturing, and selling high-quality products for household, commercial, and industrial use. This session has been called by Mel to determine where to cut 12 percent

The names and positions of the team members are:

- Mel, president of the restaurant kitchen products division
- Tom, senior vice president, engineering
- Kate, vice president, Six Sigma and quality
- Bill, senior vice president, manufacturing
- Andrew, vice president, finance
- Joanne, senior vice president, sales and marketing;
- Deonne, vice president, human resources

out of his division's budget, as the corporation has mandated for all six of its divisions.

As you read the case, pay particular attention to how Mel operates as a collaborative leader—listening, mentally processing, waiting, not trying to control matters, yet constantly working to build and sustain team unity and collective wisdom while still making his viewpoints known. Be alert to how he weaves in the elements of the integrative framework for building collaborative partnerships from Chapter Three. Also keep an eye on Mel's teammates and try to identify good examples of collaborative facilitation where they help the group process move ahead, as well as examples where they hinder and disrupt the group process. This is not a simple, straightforward example of collaborative leadership; Mel will be tested.

My comments down the right-hand side of the page are intended as a bridge to the key learning points in the previous chapters. For your first reading of the case, you may want to cover my comments and immerse yourself by making your own notes at various points throughout the dialogue. Have fun with this case. Pretend you are sitting in a corner watching the group operate. What do you see? What feedback would you give the group members regarding the things they did well? In order to improve, what would you challenge them to do differently next time?

THE BUDGET CUT SESSION

Mel: Okay. The one item on our agenda today is going to be the budget cut that Corporate has mandated for all divisions. Our desired outcome is: "Initial set of budget cut recommendations developed." We have 90 minutes, 1:00 to 2:30, to work this agenda topic. Andrew is going to review the specifics and bring us up-to-date on the details, then we will start looking at where we might make our cuts. Andrew . . .

Clear-cut agenda setting; brings in Andrew to kick off the session.

Andrew: Thanks, Mel. Well, I sent this e-mail out to all of you yesterday afternoon which stated that Corporate has mandated a 12 percent budget cut for our division. I also stated in the e-mail that the budget cut is 12 percent across the board. In other words, to be shared by each of our . . .

Andrew starts to review the facts of the situation but doesn't get very far.

Bill (*irate*): I got your e-mail, Andrew, and I read it. I understand the point about 12 percent across the board, but I don't know where you from finance get

Bill gate-closes Andrew with a personal attack; emotions are high right from the start.

off telling me that I have to reduce my budget by 12 percent!

Andrew (*heatedly*): Well, I believe it's my job!

Andrew wards off the attack with an angry defensive rebuttal. The lid is blowing off.

Mel: (firm but calm) Bill, you're attacking the messenger. Andrew didn't cut the 12 percent, he's just bringing you the information. He's innocent. It came to him as an action item from the corporatewide financial meeting he attended yesterday morning at headquarters. Let's stick to the issue.

Mel steps in with a nice piece of facilitation that protects Andrew and stops this personal attack/defend spiral from escalating, which was initiated by Bill immediately "shooting a messenger" bearing bad news.

Bill (*still fuming*): Okay. That's fine. If Corporate wants to take 12 percent out of this division's budget, that's fine. But you will all certainly have to agree that it cannot come out of manufacturing. If we don't get those products out the door, we don't make any money, and besides . . .

Bill shifts from attacking Andrew to emotionally venting about the issue. Defending his turf.

Kate (*annoyed and with sarcasm*): Aha, but tell me how I can cut back on our Six Sigma and quality initiatives. What kind of signal

Kate gate-closes Bill's venting to get her emotional reactions quickly into the mix. Defending her turf.

does that send to the marketplace and our customers? Save short term, pay long term; it's the same old story around here! Cut my budget and I'm telling you our customers will go away in droves as we toss junk into the marketplace.

Tom (*aggravated*): Andrew, another thing to remember is, I just took an over 12 percent budget hit when we cancelled the Daisymaid project last month. You've heard the old saying (raising his eyebrows) "I've already given at the office."?

Tom is not as emotional as he is "cutting" in his response. A grandstanding power play here to make clear he is above it all while defending his turf.

Andrew says nothing but his facial expression and body language indicate frustration and anger.

The smart thing to do, allow the venting to take place without pouring gasoline on the emotional fire by defending and attacking.

Joanne (*irritated*): Ther're two sides to that coin, Tom. What about all the fewer expenses you're going to have . . . but be that as is it may, Andrew, I want you to know that I'm going to cooperate. (Sarcastically) No more gas in the cars. Let a couple more salespeople go. Cut out coffee . . .

Gives Tom a jab, then gets sarcastic with Andrew. Defends her turf with hyperbole and biting mockery. These teammates are operating more like inmates.

Andrew (*peevishly showing his frustration*): If that's what you have to do.

Andrew can't take it anymore and starts his own "put-down" of Joanne.

Joanne (*derisively*): I'm sorry, Andrew. I really would like to be reasonable, but there's just no way! If we don't sell the products, we don't eat. (Waving her hands.) That's it.

Joanne slams the gate and interrupts Andrew to finish her venting. No one is listening, everyone is interrupting and saying, "Hell no, I won't go." As you'll see, allowing venting is a key to moving ahead.

Deonne (*composed*): Well, from my point of view, of course, I like to think of HR as a team player. And last month we affirmed that our people were very important to our company, and we worked hard to redeploy nine engineers from the Daisymaid project with the least amount of pain in their lives and the least amount of impact to our bottom line. (Now becoming worked up.) However, with all of the downsizing and reorganizations, the changes in our salesforce, with all of the skills training you all want, well, you all want all of these services, but I can't provide them for free!

Deonne comes into this fray as a cooler head making some good, rational points. But the more she talks, the more reactive and emotional she becomes as she pleads her case.

Andrew (*irritated and heated*): As I stated in my

We're back to personal attacks. Mel knows that's a

e-mail, Deonne—if you had read it and if you had listened to me earlier—I'm not asking any one particular function to cut its budget by 12 percent. It's a task that we all must share to some degree or other. I've already started making some cutbacks in my own function! (With emphasis.) That updated Financial Planning System I was going to install—I've cancelled it!

signal that venting has gone on long enough. He was right to let the team members "get it off their chests" because a key behavioral principle is this simple sequence: "feelings" "facts," "solutions." Letting off steam makes it easier to work rationally later on. Feelings color facts and data and the ability to process them logically; logic loses to emotion every time.

Budget cuts are an emotional exercise and tend to bring out the worst in team relationships as everyone selfishly attempts to protect their own "turf" without regard for the greater good of the team. The degree of collaboration at this point among these team members is nil.

Given a situation like this, it would be easy for Mel—out of utter frustration—to call a halt to the proceedings, make the cuts himself, and then tell each person how much he took out of the budget to help meet the division's Corporate mandate. And that would be the end of that.

However, no one would be happy with that autocratic process. Charges of favoritism certainly could prevail among some members harming mutual trust, both laterally with team members and vertically with Mel. Team unity would be imperiled and Mel's ability to create and maintain future collaborative efforts and partnerships with this team would be jeopardized.

So instead Mel let the venting play out—all members got their "emotional say." But with personal attacking starting up again, and more most likely coming, it's now time to reorient the team and get down to the business at hand.

Mel (*composed/unruffled, but firm*): Well, let me be clear because we have a very interesting proposition here. The 12 percent budget cut is not negotiable. We have to pay it. You don't have any choice in that decision. We will donate 12 percent! Collectively, what we do have the opportunity to do is to negotiate among ourselves as to who pays what portion for what reason. But as far as the budget cut is concerned, this division must give up 12 percent by some method or some means. If you want to have input and develop a fair process for determining who pays how much, that's fine. But it's not negotiable as to whether we are going to have it. We are going to have that much of a budget cut.

First Mel remained calm but firm. He made it clear what the discussion was about (developing a fair process to determine who pays, what portion of the budget cut for what reason) and what the discussion was not about (working to avoid a budget cut—that was not negotiable, there would be a 12 percent cut). Second, Mel firmly staked out the boundaries that delineated the team's range of freedom. He made sure the team knew it had an open door for creating its own destiny regarding the budget cuts. It was an opportunity to collaborate, to make commitments among themselves, to take charge, and to build mutual trust. Mel was being the epitome of a collaborative leader.

Bill (*taking charge*): Given the fact that we have to do this—and we've all heard the rumors that the corporation is in a revenue crunch at this time—I would just as soon have a voice in the process so that whatever budget cuts are made, are made in a fair way.

Bill gets Mel's point and quickly seizes the opening to get the team involved in developing a fair process for making the cuts. Turning a hostile and emotional relationship into a collaborative one takes place one or two people at a time. There is still plenty of work to do.

Mel: Okay. Do the rest of you feel that way?	*Checking with others to gauge the team's enthusiasm for the effort ahead.*
Tom (*with mild frustration and anger*): What can I say; Corporate says cut, we cut.	*A statement of resignation, not opposition.*
Mel: Well, why don't we look at this together?	*Encouraging team collaboration.*
Tom (*skeptically*): I think we're too divided to work this out among ourselves.	*Giving his opinion regarding his reservations. Still skeptical.*
Kate: I'm with you on that.	*Support for Tom's reservations.*
Bill (*firmly*): Wait, hold on! Let's not quit before we even try. Let me get some ideas here. (Goes to a flip chart and picks up a marker.) Process suggestions? Kate?	*Bill is dogged in his determination to move ahead. He makes a self-authorized decision to initiate a brainstorming exercise. Mel opened the door for action, Bill takes it!*
Kate: I say Mel decides and makes all the cuts himself. (Kate's idea is noted by Bill.)	*Proposing.*
Bill: Okay, got it. Let's go around the table. Deonne?	*Confirming. Structuring the exercising to be a round-robin format. Gate-opening.*
Deonne: Well, we could each submit a short report to each other presenting	*Proposing.*

our situations by outlining our current tasks in priority order, each with associated budget allocations, actual year-to-date spending, and our current outlooks to year end. (Bill scribes Deonne's idea.)	
Bill: All right. Tom?	*Gate-opening.*
Tom: Maybe we could do an analysis of our contract help and cut costs that way.	*Proposing.*
(*Bill writes Tom's idea down and nods to Joanne.*)	*(Bill, nonverbal gate-opening.)*
Joanne: Mel decides on all cuts, but after a full team discussion and a set of FYI suggestions from us.	*Proposing.*
(*Bill writes Joanne's idea down and then looks at Andrew.*)	*(Bill, nonverbal gate-opening.)*
Andrew: Do it the simplest way, an equal 12 percent cut across the board with each function determining where to slash its own 12 percent. (Bill notes it.)	*Proposing.*
Bill: Okay. I have one: eliminate excess facilities space. (*Writes own thought*	*As scribe, Bill gate-opens for himself so he gets his idea recorded.*

on flip chart.) Mel, do you have one?

Mel: Well, the truth of the matter is that I don't have one and it doesn't make sense for me to put one up for two reasons. I'm absolutely indifferent among all of them that are up there. As head of this division, my job is to meet the Corporate mandate. The money and the responsibility for running your functions have been assigned to each of you. You are each accountable to me for coming in on budget by year end regardless of how your budgets change during the year. I'm absolutely neutral as to how you meet the division's 12 percent mandate is achieved, just as long as you are comfortable with the process for getting there and it makes sense in terms of our overall operation.

Giving information. Mel demonstrates he is far more concerned with building a collaborative partnership among the team members than being a directive leader. He reminds the team that he is giving them a chance to unite in figuring out a fair process for determining how and where the budget cuts get made. But he underscores that whatever they decide has to meet the criteria of making sense for the whole operation (the greater good dimension) and they will be held accountable for meeting the new budget numbers whatever they turn out to be (the accountability element).

Bill: Okay, fair enough. Kate do you have another idea?

Agreeing. Gate-opening.

Kate: No. I'm going to pass.

Giving information.

Bill: Deonne?

Gate-opening.

Deonne: No, nothing specifically comes to mind right now.	*Giving information.*
Joanne: Flip a coin!	*Proposing.*
Bill: Flip a coin? (Laughs.) Okay, since we're brainstorming and flipping a coin is an option, it goes on our list. (Bill puts it on the flip chart list.) Anybody else have any other suggestions?	*Confirming. Maintaining process discipline. General address gate-opening.*
Tom: What about everyone giving up 6 percent of their budget across the board and then we collaborate on where the other 6 percent comes from? (Bill notes Tom's idea.)	*Proposing. That's the second one from the most skeptical person. A good example of turning around an attitude with authentic communication and genuine collaboration.*
Bill: Anyone else?	*General address gate-opening.*
Group (*in chorus*): No. Not really. I'm set. I'm fine. We have enough ideas.	*Clarifying.*
Bill: Let's see. We have a total of eight items up on the chart.	*Summarizing the current situation.*
Mel: Bill, that was good work. You brought order out of chaos and got us on a collaborative track. Thanks.	*Summarizing/encouraging.*

Bill: Glad to do it. (*Bill sits down.*)

Giving information.

Mel: What I'd like to do now is hold an open discussion on each of the eight ideas and build a "pros and cons" list for each item. While some ideas are more narrow and specific, they all are discrete, so we can stick with the original list. Once we've discussed the items, we can use our criteria

Taking charge; process structuring.

It's a different meeting now. The team has transitioned out of "feelings" into the "facts" stage. Everyone is engaged and working together. It has naturally and smoothly evolved into utilizing the collaborative problem solving process, even though no one explicitly said to use it. That's the value of all the tools and processes in this book. Once team members understand them, and regard them as the core methods of team collaboration, they get applied routinely as needed on a timely basis.

Notice too how Mel resisted becoming an autocratic leader in this meeting. He easily could have taken over; instead he remained in the collaborative leadership role, advocating and promoting a collaborative partnership that he knows will greatly increase the acceptance of, and commitment to, the final decision, regardless of what it turns out to be. Mel did take charge of the process in one area by stating how he wants to proceed with the next couple of steps to make sure the review of the list of ideas and the selection of solutions are done well.

---

matrix and see which ones rise to the top as the best solutions to pursue further.

Deonne: I'll be scribe for this portion of the discussion.

Giving information; demonstrating engagement in the collaborative process.

Mel: Thanks, for your help.

Supporting.

The team spends the next 40 minutes building and discussing a "pros and cons" list for each of the eight ideas, and then uses a solution prioritization matrix to sort out the potential solutions. (Sticking with his previously stated intent, Mel, in this instance, did not vote. In other situations, it may be very appropriate for the leader to be involved in the voting.) The criteria matrix exercise produces three ideas that are well above the other five.

Total team engagement in the collaborative problem solving process.

Choice 1 was item two: a short report done by each person outlining current tasks in priority order, each with associated budget allocations, actual year-to-date spending, and current outlooks to year end (*306 points*).

Choice 2 was item eight: everyone giving up 6 percent of their budget across the board and then collaborating on where the other 6 percent comes from (*288 points*).

Choice 3 was item four: Mel decides on all cuts, but after a full team discussion and listening to a set of FYI team suggestions (*252 points*).

We'll rejoin the team as it is debating and discussing the merits and drawbacks of the top three items . . .

Kate: What about this? What if we combined the main ideas we have noted up there. Maybe there's some kind of thread there we can use to weave a solution among all three.

Proposing. An excellent proposal in the sense she is looking to meld several ideas into one. This will greatly increase the chances for a teamwide win-win if the proposal can be understood and refined.

Mel: What did you have in mind?

Encouraging; seeking out her thoughts; involving her to clarify her rough ideas.

Kate: Maybe we could agree that each of us takes a 6 percent cut right off the top as item eight pro-

Kate provides a solid explanation of her thinking to the team. The ideas—and the eventual acceptance and

poses. Then as item two suggests, we go and do a detailed analysis of our own functions and write up our results in a short report and circulate those reports. We then all meet, without Mel, and based on our submitted reports we figure out how to remove another 4 percent from our pool of budgets to get that much closer to fulfilling the mandate. We present our final results to Mel to explain where the first 10 percent of the total cuts are coming from. We also give him our ideas where to get the other 2 percent. But these are just suggestions. Mel can make up the remaining 2 percent of the division's cuts from anyone in any way he sees fit. I think that Mel has the expertise and big picture insight to help us make cuts we might be avoiding or reluctant to do. What do you all think?

commitment to them—are coming from the team, not Mel. Collaborative leadership is producing a collaborative team partnership.

Andrew: You may be on to something.

Highly encouraging.

Joanne: Let me understand: We independently determine how to cut an initial

Testing comprehension.

6 percent from each of our functions?

Kate: Right.

Confirming.

Joanne: Then we submit an analysis of our own function to each other and meet, without Mel, to figure out how take another 4 percent from our budgets. Those cuts could be uneven, some contributing more, others less, but the total has to total 4 percent more for the division. Right?

Further testing comprehension. An active behavior demonstrating listening and understanding. The team members, behaviorally, are a far cry from where they were when this topic was first brought up by Mel and Andrew. They are pulling together instead of pulling apart.

Kate: That's how I see it. Then, in addition, we suggest where another 2 percent might come from—but it is only a suggestion—Mel has the opportunity to cut the final 2 percent from anywhere at his discretion.

Kate confirms that Joanne's testing of her comprehension is correct. Kate goes on to reiterate her idea of giving Mel discretionary authority over the final 2 percent.

Joanne (*thinking about it*): Uhmmmm.

Thoughtful consideration.

Deonne: That certainly involves everyone in spreading the pain around.

Giving information; supporting by citing a key benefit of Kate's proposal.

Tom: When we meet, what happens if we can agree only on 3 percent of the cuts and have no suggestions for Mel on anything else?

Seeking clarification on a key issue.

Bill: I'd say whatever percentage remains goes to Mel on top of his 2 percent, and that total percentage all becomes his call. In your example, Mel then has 3 percent discretionary authority.	*Giving his viewpoint in the form of a proposal coupled with a clarification of Tom's example.*
Group: (*Everyone nods, vocalizes support.*)	*Unanimous agreement on Bill's proposal.*
Kate: Have we covered all the bases on this?	*Seeking information, asking if there are loose ends to be discussed.*
Mel: I wonder if we have a consensus on Kate's combined proposal.	*Mel goes even a step further by seeking to test for consensus.*
Tom: Mel, could you summarize the proposal before we test for consensus to make sure we are all on the same page?	*Excellent facilitation by Tom asking Mel to summarize the proposal so everyone knows exactly what decision is being considered.*
Mel: Sure. It seems to me that what I hear the proposal being starts out with we would rather have a voice in the decision about setting up a fair process than to have everything be done unilaterally. After looking at the various alternatives, the process you're most comfortable with is: first, have every-	*A superb summary by Mel of the combined proposal laying it out as a four-step process. This clarifies exactly what the solution set is that the team will be seeking consensus on. A critical task before any consensus-seeking activity.*

one contribute one half the required cuts evenly across the board; second, have each of you do an internal functional analysis and circulate your analytical reports to each other ahead of a meeting that you will hold without me to take another 4 percent out of the budget and to outline some nonbinding suggestions for me for the final 2 percent; third, at a follow-up meeting you will inform me of your final decisions on the specifics of each function's original 6 percent, the details of the 4 percent, and the nonbinding suggestions on the 2 percent so I know where things stand; and fourth, I have the authority to make the final 2 percent of the cuts at my discretion—with that percentage maybe being slightly higher depending on the outcome of your session.

Andrew: That would get the 12 percent I need in order to prepare and send a revised divisional budget to Corporate by the end of this month.

Support/encouragement from the person at the center of the task assignment and the divisional liaison with Corporate.

Tom: Good point. Okay then, let's see what everyone thinks by testing for consensus.	*Agreement. Recommends starting the explicit consensus testing.*
Mel: Joanne, how do you feel about our current proposal?	*Mel kicks it off seeking Joanne's input.*
Joanne: I'm for it. I think it could work.	*Agreement.*
Mel: Tom?	*Gate-opening.*
Tom: I'll support it.	*Support.*
Mel: Andrew?	*Gate-opening.*
Andrew: I'm good; it fully meets my needs.	*Full agreement.*
Mel: Bill you were very concerned about being involved in the process. How do you feel about things now?	*Gate-opening; seeking information from the person most concerned with developing a fair, collaborative process.*
Bill: I like this proposal because it guarantees we all have an equal amount of skin in the game for half the cuts. It gives everybody a fair chance to be heard on their functional issues and allows us to collaborate in sorting out the next 4 percent. Finally, it gives you personal control over the final 2 percent and the	*Bill gives positive testimony and states his full agreement.*

ability to provide binding arbitration on any remaining percentages we can't or won't make as a team. I'm in full agreement.	
Mel: Deonne?	*Gate-opening.*
Deonne: Yes. I support it.	*Support.*
Mel: Kate, you did the heavy lifting in building the original combined proposal, where do you stand now?	*Crediting Kate and opening the gate for her input.*
Kate (*laughs*): It's terrific; I'm 100 percent for it.	*Agreement. Well done Kate!*
Mel: Okay. It sounds very much to me like we have a consensus. This was a tough one. We didn't hit our originally stated desired outcome but actually wound up in a better place. Once we got over the hump, we worked collaboratively through this critical and emotional business issue and we reached consensus on a sound process to stay united and to keep moving forward. Nice job! Finally, in the spirit of continuous meeting improvement, here is our meeting assessment. Complete it and give it to me as you leave. Thanks.	*Definitely the team has reached consensus. The explicit test of consensus for the four-step proposal by going around the table revealed "four agreements" and "two supports". A win-win collaborative outcome has been achieved with full commitment and acceptance by every team member. There is no "back door" through which anyone can retreat. There is still work to be done, and it may be bumpy, but the collaborative partnership for this task has been achieved and mutual trust built. It was collaborative leadership in action.*

Managing conflict and strong feelings on the road to consensus is rarely smooth or straight. As this case illustrates, it requires a united effort, orchestrated by a leader with collaborative leadership understanding, to work through tough, emotional issues. Mel, acting as a facilitative leader—in combination with other group members performing as secondary facilitators—practiced a number of tools, techniques, and processes espoused in this book. Doing so enabled them to turn an explosive situation into one that reached a productive outcome.

With Mel's guidance, the team was able to discover that the true source of conflict and feelings was its collective failure to initially understand and accept the non negotiable fact that the division was going to have a 12 percent budget cut and that each vice president was going to have to ante up a chunk of that percentage. Once that became clear and it sank in, the team collaborated on achieving a win-win solution.

Take a look back at Figure 3-1, in particular the six ingredients for collaborative partnerships, and you'll see that they all were present in this situation, with Mel verbalizing a few of them to keep the team united and moving forward.

Also, notice the leadership behaviors in Figure 3-1. Mel built mutual trust by engaging everyone and getting commitments made. He refused to be an autocratic decision maker; instead, he practiced whole group, win-win, consensus decision making (WG-I). Mel delegated to the team the action for building a fair process for determining who would pay, what portion of the total budget cut, and he stayed away from being a vocal advocate for a solution. Instead Mel facilitated the team members in uncovering a solution for themselves since they were the true stakeholders in whatever decision would eventually be made. He resolved conflict by expertly moving the team through the three stages of the feelings-facts-solutions sequence, and by clarifying and enforcing what the decision outcome "was" and "was not" about. Finally, he let the team slide into using aspects of the problem solving process, and he requested that several specific tools and processes be applied to help ensure the solution result was of the highest quality.

MY GIFT TO YOU

If you have read this far, then a gift, a reward, for your effort is due. I am giving you a *Round Tuit*! Guard it with your life, as Tuits are very hard to come by—especially the round ones. My gift is an indispensable one. It will help you become a collaborative leader—a builder of united judgment and collaborative partnerships. For years we have heard people say, "I'll do it when I get a Round Tuit." Now that you have one, you can put into practice what has been written about here. You cannot—at least with these tools and processes—set them aside and use that old excuse, "I'll start using and practicing them when I get a Round Tuit" *Now you have one, so get busy becoming a collaborative leader!!*

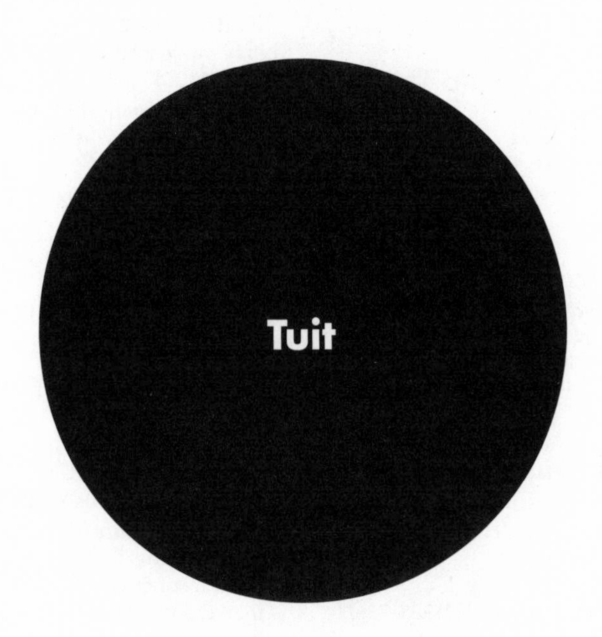

INDEX